The IMA Volumes in Mathematics and its Applications

Volume 44

Series Editors
Avner Friedman   Willard Miller, Jr.

Institute for Mathematics and its Applications
IMA

The **Institute for Mathematics and its Applications** was established by a grant from the National Science Foundation to the University of Minnesota in 1982. The IMA seeks to encourage the development and study of fresh mathematical concepts and questions of concern to the other sciences by bringing together mathematicians and scientists from diverse fields in an atmosphere that will stimulate discussion and collaboration.

The IMA Volumes are intended to involve the broader scientific community in this process.

Avner Friedman, Director
Willard Miller, Jr., Associate Director

* * * * * * * * * *

IMA ANNUAL PROGRAMS

1982–1983	**Statistical and Continuum Approaches to Phase Transition**
1983–1984	**Mathematical Models for the Economics of Decentralized Resource Allocation**
1984–1985	**Continuum Physics and Partial Differential Equations**
1985–1986	**Stochastic Differential Equations and Their Applications**
1986–1987	**Scientific Computation**
1987–1988	**Applied Combinatorics**
1988–1989	**Nonlinear Waves**
1989–1990	**Dynamical Systems and Their Applications**
1990–1991	**Phase Transitions and Free Boundaries**
1991–1992	**Applied Linear Algebra**
1992–1993	**Control Theory and its Applications**
1993–1994	**Emerging Applications of Probability**

IMA SUMMER PROGRAMS

1987	**Robotics**
1988	**Signal Processing**
1989	**Robustness, Diagnostics, Computing and Graphics in Statistics**
1990	**Radar and Sonar**
1990	**Time Series**
1991	**Semiconductors**
1992	**Environmental Studies: Mathematical, Computational, and Statistical Analysis**

* * * * * * * * * *

SPRINGER LECTURE NOTES FROM THE IMA:

The Mathematics and Physics of Disordered Media

Editors: Barry Hughes and Barry Ninham
(Lecture Notes in Math., Volume 1035, 1983)

Orienting Polymers

Editor: J.L. Ericksen
(Lecture Notes in Math., Volume 1063, 1984)

New Perspectives in Thermodynamics

Editor: James Serrin
(Springer-Verlag, 1986)

Models of Economic Dynamics

Editor: Hugo Sonnenschein
(Lecture Notes in Econ., Volume 264, 1986)

Richard McGehee Kenneth R. Meyer
Editors

Twist Mappings and Their Applications

With 41 Illustrations

Springer-Verlag
New York Berlin Heidelberg London Paris
Tokyo Hong Kong Barcelona Budapest

Richard McGehee
School of Mathematical Sciences
University of Minnesota
Minneapolis, MN 55455
USA

Kenneth R. Meyer
Department of Mathematical Sciences
University of Cincinnati
Cincinnati, OH 45221
USA

Series Editors:
Avner Friedman
Willard Miller, Jr.
Institute for Mathematics and its Applications
University of Minnesota
Minneapolis, MN 55455
USA

Mathematics Subject Classifications (1991): 58-06, 58F05, 34C35, 70H99

Library of Congress Cataloging-in-Publication Data
McGehee, Richard.
Twist mappings and their applications / Richard McGehee, Kenneth R. Meyer, editors.
p. cm. – (IMA volumes in mathematics and its applications; v. 44)
Includes bibliographical references.
ISBN 0-387-97858-5. – ISBN 3-540-97858-5
1. Twist mappings (Mathematics) I. Meyer, Kenneth R. (Kenneth Ray), 1937- . II. Title. III. Series.
QA614.8.M35 1992
515′.352–dc20 92-14149

Printed on acid-free paper.

Production managed by Hal Henglein; manufacturing supervised by Jacqui Ashri.
Camera-ready copy prepared by the IMA.
Printed and bound by Edwards Brothers, Inc., Ann Arbor, MI.
Printed in the United States of America.

9 8 7 6 5 4 3 2 1

ISBN 0-387-97858-5 Springer-Verlag New York Berlin Heidelberg
ISBN 3-540-97858-5 Springer-Verlag Berlin Heidelberg New York

The IMA Volumes
in Mathematics and its Applications

Current Volumes:

Volume 1: Homogenization and Effective Moduli of Materials and Media
Editors: Jerry Ericksen, David Kinderlehrer, Robert Kohn, J.-L. Lions

Volume 2: Oscillation Theory, Computation, and Methods of Compensated Compactness
Editors: Constantine Dafermos, Jerry Ericksen,
David Kinderlehrer, Marshall Slemrod

Volume 3: Metastability and Incompletely Posed Problems
Editors: Stuart Antman, Jerry Ericksen, David Kinderlehrer, Ingo Muller

Volume 4: Dynamical Problems in Continuum Physics
Editors: Jerry Bona, Constantine Dafermos, Jerry Ericksen,
David Kinderlehrer

Volume 5: Theory and Applications of Liquid Crystals
Editors: Jerry Ericksen and David Kinderlehrer

Volume 6: Amorphous Polymers and Non-Newtonian Fluids
Editors: Constantine Dafermos, Jerry Ericksen, David Kinderlehrer

Volume 7: Random Media
Editor: George Papanicolaou

Volume 8: Percolation Theory and Ergodic Theory of Infinite Particle Systems
Editor: Harry Kesten

Volume 9: Hydrodynamic Behavior and Interacting Particle Systems
Editor: George Papanicolaou

Volume 10: Stochastic Differential Systems, Stochastic Control Theory and Applications
Editors: Wendell Fleming and Pierre-Louis Lions

Volume 11: Numerical Simulation in Oil Recovery
Editor: Mary Fanett Wheeler

Volume 12: Computational Fluid Dynamics and Reacting Gas Flows
Editors: Bjorn Engquist, M. Luskin, Andrew Majda

Volume 13: Numerical Algorithms for Parallel Computer Architectures
Editor: Martin H. Schultz

Volume 14: Mathematical Aspects of Scientific Software
Editor: J.R. Rice

Volume 15: Mathematical Frontiers in Computational Chemical Physics
Editor: D. Truhlar

Volume 16: Mathematics in Industrial Problems
by Avner Friedman

Volume 17: Applications of Combinatorics and Graph Theory to the Biological and Social Sciences
Editor: Fred Roberts

Volume 18: q-Series and Partitions
Editor: Dennis Stanton

Volume 19: Invariant Theory and Tableaux
Editor: Dennis Stanton

Volume 20: Coding Theory and Design Theory Part I: Coding Theory
Editor: Dijen Ray-Chaudhuri

Volume 21: Coding Theory and Design Theory Part II: Design Theory
Editor: Dijen Ray-Chaudhuri

Volume 22: Signal Processing: Part I - Signal Processing Theory
Editors: L. Auslander, F.A. Grünbaum, J.W. Helton, T. Kailath, P. Khargonekar and S. Mitter

Volume 23: Signal Processing: Part II - Control Theory and Applications of Signal Processing
Editors: L. Auslander, F.A. Grünbaum, J.W. Helton, T. Kailath, P. Khargonekar and S. Mitter

Volume 24: Mathematics in Industrial Problems, Part 2
by Avner Friedman

Volume 25: Solitons in Physics, Mathematics, and Nonlinear Optics
Editors: Peter J. Olver and David H. Sattinger

Volume 26: Two Phase Flows and Waves
Editors: Daniel D. Joseph and David G. Schaeffer

Volume 27: Nonlinear Evolution Equations that Change Type
Editors: Barbara Lee Keyfitz and Michael Shearer

Volume 28: Computer Aided Proofs in Analysis
Editors: Kenneth Meyer and Dieter Schmidt

Volume 29: Multidimensional Hyperbolic Problems and Computations
Editors: Andrew Majda and Jim Glimm

Volume 30: Microlocal Analysis and Nonlinear Waves
Editors: Michael Beals, R. Melrose and J. Rauch

Volume 31: Mathematics in Industrial Problems, Part 3
by Avner Friedman

Volume 32: Radar and Sonar, Part I
by Richard Blahut, Willard Miller, Jr. and Calvin Wilcox

Volume 33: Directions in Robust Statistics and Diagnostics: Part I
Editors: Werner A. Stahel and Sanford Weisberg

Volume 34: Directions in Robust Statistics and Diagnostics: Part II
Editors: Werner A. Stahel and Sanford Weisberg

Volume 35: Dynamical Issues in Combustion Theory
Editors: P. Fife, A. Liñán and F.A. Williams

Volume 36: Computing and Graphics in Statistics
Editors: Andreas Buja and Paul Tukey

Volume 37: Patterns and Dynamics in Reactive Media
Editors: Harry Swinney, Gus Aris and Don Aronson

Volume 38: Mathematics in Industrial Problems, Part 4
by Avner Friedman

Volume 39: Radar and Sonar, Part II
Editors: F. Alberto Grünbaum, Marvin Bernfeld and Richard E. Blahut

Volume 40: Nonlinear Phenomena in Atmospheric and Oceanic Sciences
Editors: G.F. Carnevale and R.T. Pierrehumbert

Volume 41: Chaotic Processes in the Geological Sciences
Editor: David Yuen

Volume 42: Partial Differential Equations with Minimal Smoothness and Applications
Editors: B. Dahlberg, E. Fabes, R. Fefferman, D. Jerison, C. Kenig and J. Pipher

Volume 43: On the Evolution of Phase Boundaries
Editors: M.E. Gurtin and G. McFadden

Volume 44: Twist Mappings and Their Applications
Editors: R. McGehee and K.R. Meyer

Forthcoming Volumes:

1989-1990: *Dynamical Systems and Their Applications*

Dynamical Theories of Turbulence in Fluid Flows

Summer Program 1990: *Time Series in Time Series Analysis*

Time Series (2 volumes)

1990-1991: *Phase Transitions and Free Boundaries*

Shock Induced Transitions and Phase Structures

Microstructure and Phase Transitions

Statistical Thermodynamics and Differential Geometry of Microstructured Material

Free Boundaries in Viscous Flows

Variational Problems

Degenerate Diffusions

Summer Program 1991: *Semiconductors*

Semiconductors (2 volumes)

1991-1992: *Phase Transitions and Free Boundaries*

Sparse Matrix Computations: Graph Theory Issues and Algorithms

Combinatorial and Graph-Theoretic Problems in Linear Algebra

FOREWORD

This IMA Volume in Mathematics and its Applications

TWIST MAPPINGS AND THEIR APPLICATIONS

is based on the proceedings of a workshop which was an integral part of the 1989-90 IMA program on "Dynamical Systems and their Applications". The workshop brought together many of the leading figures in the modern study of twist maps.

We thank Shui-Nee Chow, Martin Golubitsky, Richard McGehee, Ken Meyer, Jürgen Moser, Clark Robinson, George R. Sell, and Eduard Zehnder for organizing the meeting and, especially, Richard McGehee and Ken Meyer for editing the volume.

Avner Friedman

Willard Miller, Jr.

PREFACE

In the 1890 volume of *Acta Mathematica*, H. Poincaré published his prize-winning paper on the stability of orbits of the three body problem. In that paper, he introduced some of the basic ideas about twist maps of the annulus. One hundred years later, the study of twist maps is still an active and important area of dynamical systems theory.

This volume presents some of the recent developments in the area. These developments include a number of topics. A great deal of activity still surrounds the basic topic introduced by Poincaré: the study of area-preserving maps of the two-dimensional annulus. Many new tools have been introduced into this study; in particular, there is a strong cross-fertilization of ideas between mathematics and physics. These new ideas have inspired many researchers to study analogs of twist maps for higher-dimensional annuli. These studies in higher dimensions hold great promise for continued progress on the basic questions of stability in mechanical systems that inspired Poincaré's original work.

We would like to express our appreciation to all the lecturers, participants, and authors who contributed to the workshop and to this collection of papers. We are grateful also to the staff and directors of the IMA for providing efficiently and unobtrusively the vast amount of support necessary for the smooth operation of the workshop and for the successful completion of this volume.

Richard McGehee

Kenneth R. Meyer

CONTENTS

A REMARK ON THE TOPOLOGICAL ENTROPY AND INVARIANT CIRCLES OF AN AREA PRESERVING TWISTMAP

SIGURD B. ANGENENT*

Let A be the annulus $S^1 \times [0, 1]$, and let $f : A \to A$ be an area preserving twist homeomorphism of A. The two boundary components of A, $A_j = S^1 \times \{j\}$, are invariant under f^q, and we shall denote the rotation number of $f|A_j$ by ρ_j.

In this note we wish to point out that the following holds:

THEOREM A. *If the topological entropy $h_{top}(f)$ of f vanishes, then f must have an invariant circle of rotation number ω, for any $\omega \in (\rho_0, \rho_1)$.*

In fact, we'll show that if "one of the invariant circles of f is missing," there must exist a compact subset $K \subset A$ which is invariant under f^q, for some $q \geq 1$, and such that $f^q|K$ has a Bernoulli shift as a factor.

If the map f is a $C^{1,\epsilon}$ diffeomorphism, then a theorem of A. Katok [**K2**] implies that f must have a "horse shoe" if $h_{top}(f) > 0$. Thus our theorem says that *any $C^{1,\epsilon}$ twist diffeomorphism of the annulus either has a transversal homoclinic point, or else it has invariant circles for any rotation number in its rotation interval* (ρ_0, ρ_1). Yet another way of phrasing this result is as follows: "*Any zone of instability of a $C^{1,\epsilon}$ area preserving twistmap contains a transverse homoclinic point for the map.*"

We shall give two proofs of this theorem. The first proof consists of simply combining two results obtained by Dick Hall and Phil Boyland.

Indeed, in [**BH**] they showed that, if one of the invariant circles of f is missing, then for some p and q with $\gcd(p, q) = 1$ the map must have a periodic orbit of type (p, q) which is not a Birkhoff orbit. On the other hand, Boyland showed in [**B**] that a twistmap with a non Birkhoff periodic orbit of type (p, q) (where $\gcd(p, q) = 1$) must have positive topological entropy, which clearly implies the theorem.

Boyland's proof of the second result which we just quoted is a fine application of Thurston's classification of surface diffeomorphisms: one of the points we wish to make in this note is that one can give an "elementary" proof of theorem A. Indeed, in [**A**] we gave an alternative and self contained proof of Boyland's criterion for positive entropy. We'll show that the method of [**A**] can also be used to prove theorem A.

We begin the proof by recalling some of the conclusions of Birkhoff's investigations on twistmaps. Birkhoff showed that any invariant circle is the graph of a Lipschitz continuous function (i.e. has the form $\{(x, \varphi(x)) : x \in S^1\}$) and that

---

*Department of Mathematics, UW - Madison. The work that led to this short note was supported by an NSF grant (number DMS - 8801486).

the Lipschitz constant of φ is bounded by a constant which only depends on the map f. To be sure, Birkhoff only proved this for C^1 maps, but as A. Katok observed in **[K1]** the result is also true for twist homeomorphisms, if one replaces "Lipschitz continuous" by "continuous," and "Lipschitz constant" by "modulus of continuity." The use of this estimate is that it implies that the set of invariant circles (and hence the set of rotation numbers which can occur) is closed.

If we assume that at least one invariant circle is missing, it follows that there is an entire interval (ρ_1, ρ_2) such that the map will not have an invariant circle with rotation number ρ for any $\rho \in (\rho_1, \rho_2)$. We may assume that (ρ_1, ρ_2) is a maximal interval with this property, so that there will be invariant circles $y = \varphi_k(x)$ with rotation numbers ρ_k $(k = 1, 2)$, and none for $\rho_1 < \rho < \rho_2$. The region

$$Z = \{(x, y) \mid x \in S^1, \varphi_1(x) \leq y \leq \varphi_2(x)\}$$

is what Birkhoff called a *zone of instability*. This region is homeomorphic to an annulus, e.g. via the homeomorphism $\Phi : A \to Z$ given by

$$\Phi(x, y) = (x, y\varphi_1(x) + (1 - y)\varphi_2(x)).$$

If we conjugate the twistmap f with this homeomorphism, then the result, $g = \Phi^{-1} f \Phi$, is again a twistmap, as the reader can easily verify.

We shall now forget about the original twistmap, and continue with g, which, by Birkhoff's construction, has no invariant circles at all, except the two boundary components of the annulus A.

To construct special orbits of the new map $g : A \to A$, we'll consider its generating function $h(x_0, x_1)$ (h was first introduced by Poincaré, and its definition and construction is given in Mather's paper **[M]**.) But first we observe that we may assume that g is defined on the infinite cylinder $S^1 \times \mathbf{R}$: outside of the annulus A we simply define it as

$$g(x, y) = \begin{cases} g(x, 0) + (y, 0), & \text{when } y < 0 \\ g(x, 1) + (y, 0), & \text{when } y > 1. \end{cases}$$

With this definition any circle $S^1 \times \{y\}$ with $y \geq 1$ or $y \leq 0$ is invariant under g.

Choose a lift $G : \mathbf{R}^2 \to \mathbf{R}^2$ of the map g, and let $h \in C^1(\mathbf{R}^2)$ be the generating function for G. Thus a sequence $\{x_k : k \in \mathbf{Z}\}$ is the sequence of x - coordinates of an orbit $\{(x_k, y_k) : k \in \mathbf{Z}\}$ of G if and only if it satisfies

(1) $$\Delta(x_{k-1}, x_k, x_{k+1}) = 0 \quad (\forall k \in \mathbf{Z})$$

where $\Delta(a, b, c) = h_2(a, b) + h_1(b, c)$, and the h_i denote the partial derivatives of h. The twist property of the map g guarantees that the function $\Delta(a, b, c)$ is strictly increasing in a and c, and the way we defined G outside of the annulus A is such that $\Delta(a, b, c) \to \pm\infty$ if $c \to \pm\infty$ or $a \to \pm\infty$ (this follows from the fact that G has the "infinite twist property," i.e. that $\lim_{y \to \pm\infty} pr_1(G(x, y)) = \pm\infty$.)

In [**A**] we studied recurrence relations like (1), and our main observation was that they may be considered as a discrete analog of a second order elliptic pde, meaning that one can use Perron's method of sub- and super harmonic functions to construct solutions of (1).

Thus we showed that if one defines a *sub solution* for (1) to be a sequence $\underline{x}_k$ with $\Delta(\underline{x}_{k-1}, \underline{x}_k, \underline{x}_{k+1}) \geq 0$ for all $k \in \mathbf{Z}$, and likewise a super solution to be sequence with $\Delta(\overline{x}_{k-1}, \overline{x}_k, \overline{x}_{k+1}) \leq 0$ $(\forall k \in \mathbf{Z})$, then, *given any sub solution* $\underline{x}_k$ *and super solution* $\overline{x}_k$ *which satisfy* $\underline{x}_k \leq \overline{x}_k$ $(\forall k \in \mathbf{Z})$ *there exists a solution* x_k *of* (1) *with* $\underline{x}_k \leq x_k \leq \overline{x}_k$ $(\forall k \in \mathbf{Z})$. The proof of this statement is entirely elementary.

Using this general method for constructing solutions from sub- and super solutions, we proved the following criterion for positivity of the topological entropy of the map g:

PROPOSITION. *If there exist a sub solution* $\underline{x}_k$ *and a super solution* $\overline{x}_k$ *for* (1) *with*

$$(2') \qquad \limsup_{k\to\infty} \frac{\underline{x}_k}{k} \leq \omega_0 \qquad \liminf_{k\to-\infty} \frac{\underline{x}_k}{k} \geq \omega_1$$

and

$$(2'') \qquad \limsup_{k\to-\infty} \frac{\overline{x}_k}{k} \leq \omega_0 \qquad \liminf_{k\to\infty} \frac{\overline{x}_k}{k} \geq \omega_1$$

for some $\omega_0 < \omega_1$*,then* $h_{top}(g) > 0$*. In fact, there is a compact* $K \subset A$ *which is invariant under some iterate* g^q *of* g*, such that* $g^q|K$ *has a Bernoulli shift as a factor.*

The proof is again elementary, and exploits the fact that the recurrence relation (1) is invariant under the $\mathbf{Z}^2$ action $(\tau_{m,n}x)_k = x_{k-m} + n$ on the space of biinfinite sequences $\mathbf{R}^{\mathbf{Z}}$. Using this invariance one can construct a large number of sub and super solutions by taking suprema and infima of translates $\tau_{m,n}(\underline{x})$ and $\tau_{m,n}(\overline{x})$ of the given sub and super solution. The Perron - process then provides one with a lot of solutions of (1) for which the corresponding orbits of the map g constitute the set K of the proposition.

The idea behind our proof of theorem A is the following. Assume the existence of an orbit $\{(x_n, y_n) \in S^1 \times [0,1] : n \in \mathbf{Z}\}$ of g which "connects the lower boundary A_0 of the annulus with its upper boundary A_1," in the sense that $\lim_{n\to\infty} y_n = 1$ and $\lim_{n\to-\infty} y_n = 0$. Corresponding to this orbit we have the sequence of x - coordinates $\{\underline{x}_n : n \in \mathbf{Z}\}$ of the lifted orbit $\{(\underline{x}_n, y_n) \in \mathbf{R} \times [0,1] : n \in \mathbf{Z}\}$ where $\underline{x}_n \,(\mathrm{mod}\, \mathbf{Z}) = x_n$. If we choose $\omega_0 < \omega_1$ so that the rotation number of $g|A_0$ is less than ω_0, and the rotation number of $g|A_1$ is more than ω_1, then the sequence $\{\overline{x}_n : n \in \mathbf{Z}\}$ is a solution of (1) which satisfies the inequalities (2″). If one also assumes that there is an orbit $\{(x'_n, y'_n) : n \in \mathbf{Z}\}$ going the other way (i.e. $\lim_{n\to-\infty} y'_n = 1$ and $\lim_{n\to\infty} y'_n = 0$), then its corresponding sequence of x - coordinates $\{\underline{x}_n : n \in \mathbf{Z}\}$ is a solution of (1) which satisfies the inequalities (2'). By the proposition the map g then must have positive topological entropy.

Unfortunately, it's not clear to me whether any of the two orbits

$$\{(\underline{x}_n, y_n) : n \in \mathbf{Z}\}, \qquad \{(\overline{x}_n, y_n) : n \in \mathbf{Z}\},$$

should exist. However, a theorem of Birkhoff's provides us with finite orbit segments which have approximately the same behaviour as the orbits whose existence we just presumed. Below we'll show that one can extend the x_n - sequences corresponding to Birkhoff's finite orbit segments, in such a way that they become sub- and super solutions satisfying (2), so that we can apply the proposition to complete the proof of theorem A. The way we extend the x_n - sequences follows an idea of Hall and Boyland's.

Let $[\sigma_0, \sigma_1]$ be the rotation interval of the map g on the annulus A, and choose *rational* numbers $\omega_0, \omega_1 \in (\sigma_0, \sigma_1)$ with $\omega_0 < \omega_1$. Without loss of generality we may assume that $\sigma_0 > 0$, and hence that all rotation numbers involved are positive. It follows from the work of J. N. Mather and S. Aubry that there exist Birkhoff orbits with rotation numbers ω_0 and ω_1 respectively. If we denote the corresponding sequences of x - coordinates by $\underline{z}_k$ and $\overline{z}_k$ ($k \in \mathbf{Z}$), then the fact that these sequences come from Birkhoff orbits implies that they are monotone, i.e.

$$\underline{z}_k < \underline{z}_{k+1} \qquad \overline{z}_k < \overline{z}_{k+1},$$

and that they are periodic in the sense that

$$\underline{z}_{k+q_0} \equiv \underline{z}_k + p_0, \qquad \overline{z}_{k+q_1} \equiv \overline{z}_k + p_1$$

for all $k \in \mathbf{Z}$ (where $\omega_j = p_j/q_j$ and $\gcd(p_j, q_j) = 1$).

Consider any orbit $\{(x_n, 0) : n \in \mathbf{Z}\}$ of the map G restricted to the lower boundary of the annulus. Since $G|A_0$ is a circle homeomorphism, the sequence x_n must be monotone: $x_n < x_{n+1}$ ($\forall n \in \mathbf{Z}$). Then there exists an integer n_0 (independent of the orbit under consideration), such that for some $k \in \{0, \ldots, n_0 - 1\}$ and some $l \in \mathbf{Z}$, one has

$$\underline{z}_l \le x_k < x_{k+1} \le \underline{z}_{l+1}.$$

Indeed, if this were not true, then any interval $(\underline{z}_l, \underline{z}_{l+1})$ would contain at most one x_i, which contradicts the fact that the rotation number σ_0 of $G|A_0$ is less than the rotation number ω_0 of the Birkhoff orbit corresponding to the $\underline{z}_k$'s.

By continuity there is a $\delta > 0$ such that the same holds for any orbit segment of the map G with length n_0, which stays in the (narrow) strip $S_0(\delta) = \mathbf{R} \times (0, \delta)$.

After increasing n_0, if necessary, the same arguments show that we may assume that for any orbit segment $\{(x_i, y_i) : 0 \le i \le n_0\}$ of the map G which is contained in the strip $S_1(\delta) = \mathbf{R} \times (1 - \delta, 1)$ there exist $0 \le k < n_0$ and $l \in \mathbf{Z}$ with

$$x_k \le \overline{z}_l < \overline{z}_{l+1} \le x_{k+1}.$$

Birkhoff showed that in a zone of instability for an area preserving twist map there exist orbits which stay arbitrarily long near one of the boundary components, then wander around in the annulus (in some unspecified way), and finally stay arbitrarily long near the other boundary component. More precisely, he showed that there is an orbit segment $\{(x_n, y_n) : -N \le n \le N\}$ for the map G, such that $(x_n, y_n) \in S_0(\delta)$ for $-N \le n < -N + n_0$, and $(x_n, y_n) \in S_1(\delta)$ for $N - n_0 < n \le N$.

We had chosen the neighbourhoods $S_0(\delta)$ and $S_1(\delta)$ so small and the number n_0 so large that there exist $n_- \in \{-N, \dots, -N+n_0-1\}$ and $n_+ \in \{N-n_0, \dots, N-1\}$ and also $k_\pm \in \mathbf{Z}$ such that

$$\underline{z}_{k_-} \le x_{n_-} < x_{n_-+1} \le \underline{z}_{k_-+1} \tag{3}$$

and

$$x_{n_+} \le \overline{z}_{k_+} < \overline{z}_{k_++1} \le x_{n_++1}. \tag{4}$$

Now we can define our super solution:

$$\overline{x}_j = \begin{cases} \underline{z}_{k_-+j-n_-} & \text{if } j \le n_-, \\ x_j & \text{if } n_- < j \le n_+, \\ \overline{z}_{k_+j-n_+} & \text{when } j > n_+. \end{cases}$$

For all $j \notin \{n_-, n_-+1, n_+, n_++1\}$ the sequence $\overline{x}_j$ certainly satisfies $\Delta(\overline{x}_{j-1}, \overline{x}_j, \overline{x}_{j+1}) = 0$, since $\overline{x}_j$ and $\overline{x}_{j\pm1}$ coincide with the x - coordinates of an orbit of G. For $j = n_-$ we have $\overline{x}_{j-1} = \underline{z}_{k_--1}$, $\overline{x}_j = \underline{z}_{k_-}$ and $\overline{x}_{j+1} = x_{n_-+1} \le \underline{z}_{k_-+1}$, by (3). So, since $\Delta(a, b, c)$ is a monotone increasing function of c, we find that

$$\Delta(\overline{x}_{j-1}, \overline{x}_j, \overline{x}_{j+1}) \le \Delta(\underline{z}_{k_--1}, \underline{z}_{k_-}, \underline{z}_{k_-+1}) = 0.$$

Similar arguments show that for the remaining three values of j ($n_- + 1$, n_+ and $n_+ + 1$) one also has $\Delta(\overline{x}_{j-1}, \overline{x}_j, \overline{x}_{j+1}) \le 0$, so that $\overline{x}_k$ is indeed a super solution for (1).

For large values of j the sequence coincides with either $\underline{z}_k$ or $\overline{z}_k$, so that it is clear that $\overline{x}_j$ satisfies the inequalities (2″).

We could repeat the whole procedure to produce a sub solution $\underline{x}_j$ of (1) which satisfies the inequalities (2′): together the sub and super solution $\underline{x}_j$ and $\overline{x}_j$ allow us to apply the proposition and conclude that theorem A is true.

REFERENCES

[A] S. B. Angenent, *Monotone recurrence relations, their Birkhoff orbits, and their topological entropy*, Ergodic Theory & Dynamical Systems (to appear).

[B] P. Boyland, *Braid types and a topological method of proving positive entropy*, preprint.

[BH] P. Boyland and G. R. Hall, *Invariant circles and the order structure of periodic orbits in monotone twist maps*, Topology, 26 (1987), pp. 21–35.

[K1] A. Katok, *Some remarks on Birkhoff and Mather Twist theorems*, Ergodic Theory and Dynamical Systems, 2 (1982), pp. 185–194.

[K2] A. Katok, *Lyapunov exponents, entropy and periodic orbits for diffeomorphisms*, Publications Math. I. H. E. S., 51 (1980), pp. 508–557.

[M] J. N. Mather, *Existence of Quasi-Periodic Orbits for Twist Homeomorphisms of the Annulus*, Topology, 21 (1982), pp. 457–467.

THE CONCEPT OF ANTI-INTEGRABILITY: DEFINITION, THEOREMS AND APPLICATIONS TO THE STANDARD MAP

SERGE J. AUBRY*

Abstract. By contrast with the KAM theory which essentially focuses on the perturbation of integrable (non-chaotic) dynamical systems, our approach consists in constructing a perturbation theory for fully chaotic ("anti-integrable") dynamical systems. By definition, at the anti-integrable limit of a dynamical system with discrete time, the coordinate of a trajectory at time i does not depend on its coordinate at time $i-1$ and its dynamics reduces to a Bernoulli shift.

This method is applied to the standard map which reaches its anti-integrable limit at $k=\infty$. At this limit, the chaotic trajectories are coded by arbitrary sequences of integers $\{m_i\}$. Those for which the sequence $b_i = |m_{i+1} + m_{i-1} - 2m_i|$ are bounded by some integer b, are preserved for $k > \sqrt{16+(b+2)^2\pi^2}$. This bound can be lowered for coding sequences $\{m_i\}$ fulfilling more restrictive conditions. These trajectories are exponentially unstable (with a non zero Lyapounov coefficient when it is defined) and exhibit dilating and contracting manifolds. "Exotic" trajectories such as ballistic trajectories are also proven to exist as a consequence of these theorems.

While the matrix of second order derivatives of the action of a trajectory is invertible (finite gap parameter), these trajectories are uniform functions of k down to certain critical values which can be also bounded from below. Theorems are obtained concerning the bifurcations of these chaotic trajectories and of the periodic cycles when decreasing k. As a corollary, the existence of infinitely many elliptic cycles is proven for the standard map at large k.

For large enough k, the minimal orbits of the standard map (associated with the ground-states of the Frenkel-Kontorowa model) belong to the set of anti-integrable trajectories. This property is used for proving two theorems which apply close enough to the anti-integrable limit 1-The invariant Cantori of the standard map have zero measure length 2- The left(or right) derivative of the average action of the minimal orbits as a function of their rotation number ω is a purely discrete function with all its discontinuities at the rational ω. (This last result proves partially an early conjecture which was proposed twelve years ago by the author)

Key words. Dynamical Systems, Chaos, Anti-Integrability, Structure

1. Introduction: The variational approach. My purpose in this talk, is to describe an approach which was initially developed for problems in Solid State Physics but which turned out to have strong connections with the mathematical problems concerning Hamiltonian dynamical systems. Since many of the readers of this paper should be mathematicians, I apologize in advance for mistakes in my mathematical language and for improper quotations or omissions which could be found in this work. However, I have the feeling that the approach I developed, is useful for finding new mathematical results with original methods. I included in this proceeding some new applications of the concept of anti integrability which came to my mind while attending the workshop.

Eight theorems plus some propositions and corollaries, are presented in this paper with their full proof. Four theorems are still unpublished. This paper essentially focuses on the well-known standard map introduced by H. Poincaré a century ago and which is the main topics of this workshop. However, much of the ideas presented in this work, can be generalized to wider classes of models as it has already been shown and will be shown in further publications.

*Laboratoire Léon Brillouin, CEN Saclay, F-91191-Gif-sur-Yvette, Cedex, France (Laboratoire commun CEA–CNRS)

This paper is organized in five main sections. In this first one, I present the Frenkel-Kontorowa model and the standard map. In the second section, the concept of anti-integrability and its main theorem (theorem 1) are presented. The third section describes the concept of gap parameter and the theorem (theorem 3) on the k-continuation of trajectories. Section four is devoted to the behavior of the anti-integrable trajectories at small k (theorem 5). Examples showing some possible (inverse) bifurcations of the anti-integrable trajectories, are presented for illustrations. Finally, section 5 is entirely devoted to minimal orbits in the anti-integrable regime. It mostly contains the long proof of theorem 8, which confirms an early conjecture [1, 15] about the "completeness of the Devil's Staircase" at large k.

Many structural problems in solid state physics consist in finding the absolute minimum of a variational form which is just the energy of the considered system as a function of its configuration. The initial idea of my works [1] was to associate an Hamiltonian dynamical system to a structural model. For this purpose, we just have to consider this variational form as the action of this dynamical system. Then, the space variable has to be interpreted as a time variable. However, when doing this association, the physical problem is changed. The trajectories of the dynamical system are those which extremalize the variational form (then called action) while the structures of physical interest are those which are absolute minima or at least local minima of the same variational form. Then, it is clear that one has to be careful when transposing the results obtained on the structural model to the associated dynamical system and vice versa. However, the variational approach which is rather natural for the structural problem appears as a rather new approach for the dynamical problem. Let us note, that ideas similar to this one, were proposed by I. Percival [2] (for the standard map) and in a different context by C. Radin [3] (as symbolic dynamical systems associated with classical lattice gas models).

Our work has been focused for many years [1] on the standard map, the most studied twist mapping, the action of which appears to be the energy of the Frenkel Kontorowa (FK) model which is well-known in solid state physics. This action has the form

$$\Phi(\{u_i\}) = \sum_i [W(u_{i+1} - u_i) + V(u_i)] \tag{1-a}$$

with the special choice

$$W(x) = \frac{1}{2}\, C\, x^2 - \mu.x \tag{1-b}$$

and

$$V(x) = \lambda(1 - \cos x)$$

Then, $i \in \mathbf{Z}$ is either the discrete time for the standard map, or the atomic label of the atom i (for the FK model); then $\{u_i\} \in \mathbf{R}^{\mathbf{Z}}$ represents the set of positions of the atoms i in the structure. This model can be physically represented as a chain of balls (atoms) coupled by springs (coupling constant C) and submitted to an underlying periodic potential (coupling constant λ) (cf. fig. 1)

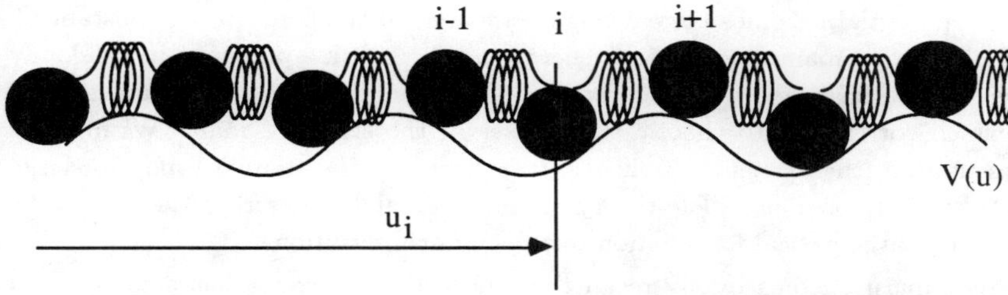

FIGURE 1: *Mechanical representation of the Frenkel-Kontorowa model (which is the Action of the Standard Map). Each ball i located at the abscissa u_i, is submitted to the periodic potential $V(u_i)$ and is elastically coupled to the two nearest neighbor balls.*

The extrema $\{u_i\}$ of this variational form (1) are equivalently the stationary configurations of the FK model or the trajectories of the standard map. Setting

(2-a) $$k = \frac{\lambda}{C}$$

the extremalization equation $\dfrac{\partial\Phi(\{u_i\})}{\partial u_i} = 0$ yields

(2-b) $$k \sin u_i + 2u_i - u_{i+1} - u_{i-1} = 0$$

By defining the conjugate momentum of u_i as

(3-a) $$p_i = u_i - u_{i-1}$$

equation (2-b) can be written again as

(3-b) $$u_{i+1} = k \sin u_i + u_i + p_i$$

(3-c) $$p_{i+1} = k \sin u_i \qquad + p_i$$

Considering u_i modulo 2π, these equations define the standard map $\mathcal{S}$

(3-d) $$\{u_{i+1}, p_{i+1}\} = \mathcal{S}\{u_i, p_i\}$$

which maps the cylinder $\mathbb{C} \times \mathbb{R} = \{u_i \mod 2\pi, p_i\}$ onto itself.

Thus by associating the FK model to the standard map $\mathcal{S}$, any trajectory of $\mathcal{S}$ can be represented as a stationary configuration of this structural model. Then, it is "physically obvious" (see fig. 1) that when the amplitude λ of the underlying periodic potential becomes large enough compared to the fixed spring constant C, the balls can be maintained by the energy barriers of this potential in randomly chosen wells. Thus, there must exist stationary configurations which are chaotic and which correspond to chaotic trajectories in the standard map. Within this representation, the existence of chaotic trajectories in the standard map for large k, is "physically obvious". The theory of anti-integrability described here will give a rigorous mathematical formulation to this physical intuition.

Preliminarily, let us briefly recall that the standard map $\mathcal{S}$ has a well-known integrable limit at $k = \lambda = 0$. Then $p_{i+1} = p_i = \cdots = p_0 = l$ is a motion invariant since it is independent of i and $\mathcal{S}$ becomes rotation with angle l for each invariant circle C_l of the cylinder $\mathbb{C}\times\mathbb{R}$ defined by $p = l = Cste$. In that simple case, the more general Kolmogorov, Arnol'd, Moser (KAM) theorems states that when $\zeta = \frac{l}{2\pi}$ is a "good" irrational number, there exists some non vanishing value $k_c(\zeta)$ of k, such that for $|k| < k_c(\zeta)$, there exists an invariant topological circle C_l on the cylinder $\mathbb{C} \times \mathbb{R}$, on which $\mathcal{S}$ is still a rotation with rotation number ζ. The trajectories $\{u_i\}$ on these $\mathcal{S}$-preserved 1-tori C_l are described by an analytic hull function $f(x; k, l)$ (which depends on k and l)

$$u_i = f(il + \alpha; k, l) = il + \alpha + g(il + \alpha; k, l) \tag{4}$$

where $g(x; k, l)$ is an analytic 2π-periodic function of x. This function $g(x; k, l)$ is also analytic with respect to k in some complex neighborhood of $k = 0$. It is not our purpose to describe here this KAM theory. Since the first KAM theorems, nearly thirty years ago, an abundant literature on this topic, has been written and new results are still obtained (for example, see this proceeding for more refined results by M. Herman [8]).

2. Anti-integrability. At the integrable limit, the behavior of a Hamiltonian dynamical system is always "smooth". There is no chaotic trajectories and although there is many exceptional trajectories like those which are periodic or those which ly on separatrices, most trajectories (with Lebesgue probability one), are quasiperiodic on invariant tori. Roughly speaking, the KAM theorems state that most trajectories which are quasi-periodic at the integrable limit, are preserved under small enough non integrable perturbation and continuously depends on the amplitude of this perturbation. The condition on the invariant tori for being preserved, is fulfilled providing "good non-resonance" conditions which occur with probability one.

By analogy and by contrast with the integrable limit, it appears interesting to study the other limits for the Hamiltonian dynamical system, where the behavior of the dynamical system would be "perfectly chaotic". Such a limit can be found for the standard map when the potential $W(x)$ in the variational form (1) vanishes. for $C = 0$ (or $k = +\infty$), the solutions of the extremalization equations $\frac{\partial\Phi(\{u_i\})}{\partial u_i} = 0$, are simply given by

$$\frac{\partial V(u_i)}{\partial u_i} = \lambda \sin u_i = 0 \tag{5-a}$$

and the whole set of stationary configurations the FK model or equivalently of trajectories of $\mathcal{S}$, is given by

(5-b) $$u_i = m_i \pi$$

where $\{m_i\} \in \mathbf{Z}^{\mathbf{Z}}$ are all possible sequences of integers. Although this limit is clearly singular since the phase space $\mathbf{C} \times \mathbf{R}$ is not filled, the dynamics of the standard map is trivially described by a Bernoulli shift. We are going to show, that under some conditions on the coding sequence $\{m_i\}$, the chaotic behavior of the trajectories at this singular limit, is preserved under small enough perturbations. It is useful to note here that the concept of anti-integrable limit is widely generalizable. For a discrete time dynamical system in $n+n$ dimensions with action

(6-a) $$\Phi(\{u_i\}) = \sum_i L(u_{i+1}, u_i)$$

where $\{u_i\} \in (\mathbf{R}^n)^{\mathbf{Z}}$ and $L(x,y)$ is a suitable function of $\mathbf{R}^n \times \mathbf{R}^n \to \mathbf{R}$, we set more generally:

DEFINITION 1. When the variational form (called action) $\Phi(\{u_i\})$ which defines a discrete time dynamical system, is separable, i.e. can be written as

(6-b) $$\Phi(\{u_i\}) = \sum_i V(u_i)$$

where $V(x)$ is a function $\mathbf{R}^n \to \mathbf{R}$, then, this dynamical system is called *anti-integrable*.

According to this definition, when the system is anti-integrable, the extremalization equation becomes $\dfrac{\partial V(u_i)}{\partial u_i} = 0$. Equation $\dfrac{\partial V(x)}{\partial x} = 0$ determines a set of extrema $E_V = \{x_\nu\}$ which are indexed by the set $\mathcal{A}$ of indices ($\nu \in \mathcal{A}$) and then, the whole set of trajectories $\{u_i\}$ is given by

(7-a) $$u_i = x_{\nu_i}$$

where $\{\nu_i\}$ are arbitrary sequences of symbols in $\mathcal{A}$

(7-b) $$\{\nu_i\} \in \mathcal{A}^{\mathbf{Z}}$$

For the above example (5), $\mathcal{A} = \mathbf{Z}$. Providing that $\mathcal{A}$ contains at least two elements, the trajectories of this map $\{u_i\} \in \mathbf{K} = E_\nu^{\mathbf{Z}}$ can be described as a Bernoulli shift. (But the phase space is not filled because u_i cannot take arbitrary values!). We now turn back to the standard map and prove under certain conditions on the coding sequence $\{m_i\}$, that these chaotic trajectories are preserved when the action at the anti-integrable limit is submitted to small enough perturbations:

THEOREM 1. *Let b be a strictly positive arbitrary integer and let us set $\mathbf{K}_b$, the set of integer sequences $\{m_i\}$ in $\mathbb{Z}^{\mathbb{Z}}$ which fulfills for all i*

$$|m_{i+1} + m_{i-1} - 2m_i| \leq b \tag{8-a}$$

Then for

$$k \geq B(b) = \sqrt{16 + (b+2)^2\pi^2} \tag{8-b}$$

there exists for each $\{m_i\}$ in $\mathbf{K}_b$, a unique trajectory $\{u_i\}$ of the standard map such that for all i

$$|u_i - m_i\pi| \leq \frac{\pi}{2} \tag{9-a}$$

This solution $u_i(k)$ continuously depends on k and we have

$$\lim_{k\to\infty} u_i(k) = m_i\pi \ . \tag{9-b}$$

By definition, the coding sequence of trajectory $\{u_i(k)\}$ is the integer sequence $\{m_i\}$.

Proof. First, for a given set of integers $\{m_i\}$ fulfilling (8-a), the domain $E\left(\{m_i\}; \frac{\pi}{2}\right)$ is defined as the subset of configurations $\{u_i\}$ which fulfill condition (9-a) for all i. The distance $d(\{u_i\},\{v_i\})$ between two configurations $\{u_i\}$ and $\{v_i\}$ in $E\left(\{m_i\}; \frac{\pi}{2}\right)$ is defined by

$$d(\{u_i\},\{v_i\}) = \operatorname*{Sup}_{i\in\mathbb{Z}} |u_i - v_i| \tag{10-a}$$

and determines the standard Banach topology.

For k large enough, we are going to define a continuous operator $\mathbf{S}$ which maps this domain $E\left(\{m_i\}; \frac{\pi}{2}\right)$ onto itself and such that its fixed points (defined by $\mathbf{S}(\{u_i\}) = \{u_i\}$) are the stationary configurations (and only them) given by eq. (2-b). Next, it will be shown that this operator is distance-contracting so that by application of the Banach fixed point theorem, the existence of a unique fixed point in $E\left(\{m_i\}; \frac{\pi}{2}\right)$ will be proven and consequently theorem 1.

For $\{u_i\} \in E\left(\{m_i\}; \frac{\pi}{2}\right)$, we define $\{v_i\} = \mathbf{S}(\{u_i\})$ by the implicit equation

$$k \sin v_i = u_{i+1} + u_{i-1} - 2u_i \tag{10-b}$$

and by the condition $\{v_i\} \in E\left(\{m_i\}; \frac{\pi}{2}\right)$ or

$$|v_i - m_i\pi| \leq \frac{\pi}{2} \qquad \text{for all } i. \tag{10-c}$$

Since for all $\{u_i\} \in E\left(\{m_i\}; \frac{\pi}{2}\right)$, we have for all i

(11-a) $$|u_{i+1} + u_{i-1} - 2u_i| \le (|m_{i+1} + m_{i-1} - 2m_i| + 2)\pi \le (b+2)\pi$$

when

(11-b) $$k > (b+2)\pi$$

the implicit equation (10-b) has one and only one solution $\{v_i\}$ in $E\left(m_i\}; \frac{\pi}{2}\right)$, and operator $\mathbf{S}$: $E\left(\{m_1\}; \frac{\pi}{2}\right) \rightarrow E\left(\{m_i\}; \frac{\pi}{2}\right)$ is well defined for all $\{u_i\} \in E\left(\{m_i\}; \frac{\pi}{2}\right)$.

The matrix elements of the Jacobian derivative $\overline{\overline{J}} = \partial\mathbf{S}$ of this operator are

(12-a) $$J_{ij} = \frac{\partial v_i}{\partial u_j} = (-1)^{m_i} \frac{1}{\sqrt{k^2 - (u_{i+1} + u_{i-1} - 2u_i)^2}} (\delta_{i,j+1} + \delta_{i,j-1} - 2\delta_{i,j})$$

($\delta_{i,j}$ is the Krönecker symbol $\delta_{i,j} = 1$ for $i = j$ and 0 otherwise). Then, using (12-a) the spectral norm of $\overline{\overline{J}}$ defined as $\|\overline{\overline{J}}\| = \operatorname{Sup}_{\overline{X}} \frac{\|\overline{\overline{J}}.\overline{X}\|}{\|\overline{X}\|}$ (with the Banach norm $\|\overline{X}\| = \operatorname{Sup}_i |X_i|$ for $\overline{X} = \{X_i\}$) is easily bounded as

(12-b) $$\|\overline{\overline{J}}\| \le \frac{4}{\sqrt{k^2 - (b+2)^2\pi^2}}$$

When $\|\overline{\overline{J}}\| < 1$, operator $\mathbf{S}$ is contracting for the Banach metric (10-a). Then, according to the Banach fixed point theorem [4], operator $\mathbf{S}$ has a unique fixed point $\{u_i\} = \mathbf{S}(\{u_i\})$ in $E\left(\{m_i\}; \frac{\pi}{2}\right)$. Condition $\|\overline{\overline{J}}\| < 1$ as well as condition (11-b) are both fulfilled when (8-b) is fulfilled.

Since operator $\mathbf{S}$ is uniformly continuous with respect to the parameter $\frac{1}{k}$ on the interval $\left[0, \frac{1}{(b+2)\pi}\right[$, a complementary theorem (ref. 4 pp. 18) allows one to prove that the fixed point $\{u_i(k)\}$ is also a continuous function of this parameter $\frac{1}{k}$ on the same interval. For $\frac{1}{k} \rightarrow 0$ or equivalently $k \rightarrow \infty$, eq. 9-b is then proven. □

This theorem predicts the existence of a subset of trajectories $\{u_i\}$ in the standard map with an arbitrarily given approximate itinerary $\{m_i\pi\}$ up to the accuracy $\frac{\pi}{2}$. Most of these trajectories are chaotic with unbounded momenta $\{p_i\} = \{u_i - u_{i-1}\}$. With the lowest possible value for b in (8-a), we find that for

(13) $$k > \sqrt{16 + 9\pi^2} \cong 10.24$$

there exists such chaotic trajectories with unbounded momenta.

However, there also exists many non chaotic trajectories. For example, it is readily proven that coding sequences $\{m_i\}$ fulfilling for all i

(14-a) $$m_{i+s} = m_i + 2r$$

where r and s are arbitrary irreducible integers, are associated for large enough k according to theorem 1, to trajectories $\{u_i\}$ which also fulfills for all i

(14-b) $$u_{i+s} = u_i + 2r\pi$$

These trajectories are periodic cycles of period s of the standard map. Other specific behavior can be obtained. For example, the choice

(14-c) $$m_i = \text{Int}\left(\frac{i^2}{2}\right)$$

yields ballistic trajectories which escape to infinity with the average acceleration $\lim_{i\to\infty} \frac{p_i}{i} = \pi$. Some of these trajectories were found by Chirikov [5], but it is clear that our construction allows one to predict the existence of infinitely many other trajectories with a non ergodic and specifically chosen behavior.

With more severe conditions on the sequence $\{m_i\}$, the lower bound in k for which there exist chaotic trajectories with unbounded momentum can be lowered. For example, we proved in ref. 6 that for the coding sequence $\{m_i\}$ which fulfills both conditions

(15-a) $$b_i = |m_{i+1} + m_{i-1} - 2m_i| \leq 1$$

and for a given integer p

(15-b) $$b_i b_{i+n} = 0 \qquad \text{for any} \quad n \leq p$$

the lower bound in k for which there exists trajectories which fulfills (9-a) is lowered. However, this trajectory is not necessarily unique in the domain $E\left(\{m_i\}; \frac{\pi}{2}\right)$. At the limit $p \to \infty$, this bound becomes

(15-c) $$k >\cong 6.86$$

For all these trajectories predicted by theorem 1, the existence of continuous dilating and contracting manifolds, have been proved [6] using a slight variation of the proof of theorem 1. Let us recall that by definition, the contracting manifold $\mathcal{V}^-(\{u_0, p_0\})$ of a trajectory $\{u_i, p_i\} = \mathcal{S}^i(\{u_0, p_0\})$ with initial point $\{u_0, p_0\}$, is the locus of points $\{v_0, q_0\}$ such that the trajectory $\{v_i, q_i\} = \mathcal{S}^i\{v_0, q_0\}$ is asymptotic to the trajectory $\{u_i, p_i\}$, that is $\lim_{i\to+\infty} |u_i - v_i| = 0$. The dilating manifold $\mathcal{V}^+(\{u_0, p_0\})$ is defined identically but for the inverse standard map, that is $\lim_{i\to-\infty} |u_i - v_i| = 0$.

3. Gap parameter, trajectory continuation. The question which comes out now is: what become at smaller k the chaotic trajectories, the existence of which is predicted by theorem 1 at large k? For answering this question, it is first useful to introduce the concept of "Gap Parameter? which is different from the concept of Lyapounov coefficient although it has some connection with it.

We consider first the Quadratic Expanded Action Matrix (QEAM) of a configuration $\{u_i\}$. (It is not required for this configuration to be a stationary configuration fulfilling (2-b)). The infinite matrix $\overline{\overline{M}}$ with elements $M_{i,j}$ $(i,j \in \mathbb{Z})$ is defined in the general case as the self-adjoint matrix of the quadratic form corresponding to the action expanded at second order for the configuration $\{u_i\}$:

$$M_{i,j} = \frac{\partial^2 \Phi(\{u_n\})}{\partial u_i \partial u_j} \tag{16-a}$$

For the associated structural model, the spectrum of this matrix just provides the phonon frequency squares. For structural problems, it is physically important to know whether zero belongs to this spectrum or not. In order to "measure" this property, we introduce the dimensionless quantity defined as

DEFINITION 2. The Gap Parameter of configuration $\{u_i\}$ is defined as

$$\Delta = \frac{1}{\|\overline{\overline{M}}\|.\|\overline{\overline{M}}^{-1}\|} \; . \tag{16-b}$$

When $\overline{\overline{M}}$ is not invertible, then $\Delta = 0$.

$\|\overline{\overline{M}}\|$ is the spectral norm of matrix $\overline{\overline{M}}$ which is now defined as $\text{Sup}_{\overline{X}} \frac{\|\overline{\overline{M}}.\overline{X}\|}{\|\overline{X}\|}$ with the Hermitian norm. For this self-adjoint operator, $\|\overline{\overline{M}}\|$ is equivalently equal to the largest eigenvalue modulus $\text{Sup}_{E_\alpha \in S(\overline{\overline{M}})} |E_\alpha|$ (where $S(\overline{\overline{M}})$ is the spectrum of $\overline{\overline{M}}$) and $\|\overline{\overline{M}}^{-1}\|$ is equal to the inverse of the smallest eigenvalue modulus. Thus, Δ is just the ratio of the smallest eigenvalue modulus and of the largest eigenvalue modulus of $\overline{\overline{M}}$. This Gap Parameter is defined for all trajectories. It is not an average quantity as the Lyapounov coefficient and might be zero while the Lyapounov coefficient be non zero (or undefined). When it is non-zero, the Lyapounov coefficient must be positive if it is defined [6,14]. However, for the particular trajectories of the standard map which are periodic cycles, it is equivalent to say that the Lyapounov coefficient is strictly positive (hyperbolic periodic cycle) or that the Gap Parameter is strictly positive (for the proof of this assertion see ref. 7 chap.6 page 138). It comes out:

THEOREM 2. *The Gap Parameter of the trajectories $\{u_n\}$ of the standard map which are obtained as attractive fixed points of the operator* **S** *defined in some* *domain* $E\left(\{m_i\}; \frac{\pi}{2}\right)$ *where* $\{m_i\}$ *fulfills (8-a) and* $k > (b+2)\pi$, *is strictly non zero:* $\Delta \neq 0$.

Proof. According to definition (16-a), the QEAM of a trajectory $\{u_n\}$ of the standard map is defined

(17-a) $$M_{i,i} = C\,(2 + k\cos u_i)$$

(17-b) $$M_{i,i+1} = M_{i,i-1} = -C$$

(17-c) $$M_{i,j} = 0 \quad \text{for } |i-j| > 1$$

Since $\{u_n\}$ is assumed to be an attractive fixed point for the operator $\mathbf{S}$ in some open domain, it fulfills the stationary equation (2-b). Then, the Jacobi matrix $\overline{\overline{J}}$ defined by (12-a) becomes at this fixed point

(18-a) $$J_{i,i} = -\frac{2}{k\cos u_i}$$

(18-b) $$J_{i,i+1} = J_{i,i-1} = \frac{1}{k\cos u_i} \quad \text{and}$$

(18-c) $$J_{i,j} = 0 \qquad \text{for } |i-j| > 1$$

The QEAM (17) can be written as

(19-a) $$\overline{\overline{M}} = C\overline{\overline{D}}(\overline{\overline{1}} - \overline{\overline{J}})$$

where $\overline{\overline{D}}$ is a diagonal matrix defined as $D_{i,j} = k\cos u_i \delta_{i,j}$. Thus, we have

(19-b) $$\|\overline{\overline{M}}\| \leq Ck(1 + \|\overline{\overline{J}}\|)\,\mathop{\mathrm{Sup}}_i |\cos u_i| \leq Ck(1 + \|\overline{\overline{J}}\|)$$

Since operator $\mathbf{S}$ is supposed to be contracting, $\|\overline{\overline{J}}\| < 1$ which implies that $(\overline{\overline{1}} - \overline{\overline{J}})$ is invertible. Using (10-b) and $k > (b+2)\pi$, it comes out

(20-a)
$$\mathop{\mathrm{Inf}}_i |\cos u_i| = \sqrt{1 - \mathop{\mathrm{Sup}}_i \sin^2 u_i}$$
$$= \sqrt{1 - \frac{1}{k^2}\ \mathop{\mathrm{Sup}}_i (u_{i+1} + u_{i-1} - 2u_i)^2} \geq c = \sqrt{1 - \frac{(b+2)^2\pi^2}{k^2}} > 0$$

Since $\mathop{\mathrm{Inf}}_i |\cos u_i| \geq c > 0$, operator $\overline{\overline{D}}$ is invertible and we have $\|\overline{\overline{D}}^{-1}\| \leq \frac{1}{kc}$. The inverse $\overline{\overline{M}}^{-1}$ can be written

(20-b) $$\overline{\overline{M}}^{-1} = \frac{1}{C}\,(\overline{\overline{1}} - \overline{\overline{J}})^{-1}\overline{\overline{D}}^{-1} = \frac{1}{C}\left(\sum_{n=0}^{\infty} \overline{\overline{J}}^n\right)\overline{\overline{D}}^{-1}$$

which yields

(20-c) $$\|\overline{\overline{M}}^{-1}\| \leq \frac{1}{C} \|\overline{\overline{D}}^{-1}\| \sum_{n=0}^{\infty} \|\overline{\overline{J}}\|^n \leq \frac{1}{C\sqrt{k^2-(b+2)^2\pi^2}} \frac{1}{1-\|\overline{\overline{J}}\|}$$

Using (19-b) and (20-c), we obtain a strictly positive lower bound for the gap parameter, which proves Proposition 1:

(20-d) $$\Delta \geq \frac{1-\|J\|}{1+\|J\|} \sqrt{1-\frac{(b+2)^2\pi^2}{k^2}} > 0 \qquad \square$$

Since this proposition proves the existence of trajectories of the standard map for which the gap parameter is non vanishing, it is now useful to show that a non vanishing gap parameter implies that the corresponding trajectories are continuous functions of the model parameters. In other words, this non trivial property means that these trajectories are not bifurcating. This proposition applies to any configuration with a non zero gap parameter and not only to those obtained from theorem 1.

THEOREM 3. *For a given $k = k_0$, let us consider a trajectory $\{u_i\}$ of the standard map with a non vanishing gap parameter. Then*

a) This trajectory $\{u_i\}$ cannot be approached uniformly by other trajectories of the standard map. (i.e. there exists a positive number η such no other trajectories $\{v_i\}$ of the standard map fulfills $|u_i - v_i| < \eta$ for all i)

b) this trajectory $\{u_i\}$ of the standard map, depends continuously and uniformly on the parameter k in some open domain in k around k_0.

Proof. Let us assume that there exists a sequence of trajectories $\{v_i^{(n)}\}$ which converges uniformly to $\{u_i\}$ for $n \to \infty$ and let us prove that this situation is impossible. By definition of the uniform convergence, $\operatorname{Sup}_i |u_i - v_i^{(n)}| = \eta_n$ and $\ll_{n\to\infty} \eta_n = 0$. Since configurations $\{u_i\}$ and $\{v_i^{(n)}\}$ both fulfills eq. 2-b, one obtains by difference

(21) $$(u_{i+1} - v_{i+1}^{(n)}) + (u_{i-1}^{(n)} - v_{i-1}^{(n)}) - 2(u_i - v_i^{(n)}) - k_0 \cos \xi_i^{(n)} (u_i - v_1^{(n)}) = 0$$

where $\xi_i^{(n)}$ is some real number which belongs to the interval determined by u_i and $v_i^{(n)}$. Thus $\ll_{n\to\infty} \operatorname{Sup}_i |u_i - \xi_i^{(n)}| = 0$ and consequently, the sequence $\{\xi_i^{(n)}\}$ converges uniformly to $\{u_i\}$. Eq. 21 shows that $\{\varepsilon_i^{(n)}\} = \{u_i - v_i^{(n)}\}$ is a bounded eigenvector of the QEAM $\overline{\overline{M}}(\{\xi_i^{(n)}\})$ of configuration $\{\xi_i^{(n)}\}$ corresponding to a zero eigenvalue. Since for $n \to \infty$, $\{\xi_i^{(n)}\}$ converges *uniformly* to $\{u_i\}$, $\overline{\overline{M}}(\{\xi_i^{(n)}\})$ converges *uniformly* to $\overline{\overline{M}}(\{u_i\})$. Then, since zero belongs to the spectrum of $\overline{\overline{M}}(\{\xi_i^{(n)}\})$, zero should also belong to the spectrum of the limit matrix $\overline{\overline{M}}(\{u_i\})$. This fact contradicts the initial assumption. As a result, the stationary configuration $\{u_i\}$ cannot be the uniform limit of stationary configurations which proves the first assertion of this proposition.

For proving the second assertion of theorem 3, let us consider the set of differential equations for $\{v_k(k)\}$, which is obtained as a formal derivation of eq. 2-b with respect to k

$$\varepsilon_{i+1}(k) + \varepsilon_{i-1}(k) - 2\varepsilon_i(k) - k \cos v_i(k).\varepsilon_i(k) = \sin\, v_i(k) \tag{22-a}$$

where $\overline{\varepsilon} = \{\varepsilon_i(k)\} = \left\{\dfrac{d\ v_i(k)}{dk}\right\}$. It can be written formally as

$$\overline{\overline{M}}(\{v_i\};k)\,\overline{\varepsilon} = C\,\overline{V} \tag{22-b}$$

where $\overline{\overline{M}}(\{v_i\};k)$ is the QEAM of trajectory $\{v_i\}$ with the parameter k defined by (22) and $\overline{V}$ is a vector with components $\{\sin\, v_i\}$. Since the gap parameter $\Delta(\{u_i\};k_0)$ of trajectory $\{u_i\}$ is assumed to be non zero, $\overline{\overline{M}}(\{u_i\};k_0)$ is invertible. Since $\overline{\overline{M}}(\{v_i\};k)$ is a uniformly continuous function of the vector $\{v_i\}$ and of the parameter k, the gap parameter $\Delta(\{v_i\};k)$ remains strictly positive for $(\{v_i\};k)$ in some neighborhood D of the point $(\{u_i\};k_0)$. In that domain, $\overline{\overline{M}}(\{v_i\};k)$ is invertible and its inverse $\overline{\overline{M}}^{-1}(\{v_i\};k)$ is a uniformly continuous function of $(\{v_i\};k)$ in some open domain D' included in D. Thus the set of differential equations

$$\frac{d\ v_i(k)}{dk} = \sum_j \overline{\overline{M}}^{-1}_{i,j}(\{v_n\};k)\ \sin\ v_j \tag{22-c}$$

can be integrated in this domain D', and for a given initial condition, yields a unique solution $\{v_i(k)\}$. Choosing the initial condition $\{v_i(k_0)\} = \{u_i\}$ for $k = k_0$, it is readily found by integration of (22-a) that this solution $\{v_i(k)\}$ fulfills eq.2-b. Thus, this solution $\{v_i(k)\}$ is a trajectory for the parameter k which depends uniformly and continuously on $k \in I$ where I is some open interval $]k_0^-, k_0^+[$ which contains k_0. (see erratum ref. 21)

□

Thus while the gap parameter does not vanish, theorems 2 and 3 show that there exists a *unique continuation* at smaller k of the "anti-integrable" trajectories obtained at large k (by theorem 1), in a parameter region where theorem 1 does not apply any more. When a value of k is reached at which the gap parameter of the trajectory vanishes, either the continuation of this trajectory is impossible or there may exist several (or infinitely many) paths for the continuation of this trajectory through bifurcations.

4. Inverse bifurcations of the anti-integrable trajectories at small k. The preliminary new results presented in this section have been suggested by useful discussions with Tom SPENCER while attending the workshop. The anti-integrable trajectories predicted by theorem 1, are characterized by their coding sequence $\{m_i\}$ which can be arbitrary providing that condition (8-a) is fulfilled for all i, for some strictly positive integer i. It is especially interesting to study the (inverse) bifurcations of these trajectories and particularly those of the periodic cycles which belong to this set, which occur by decreasing k. Studying the trajectories of the standard map by continuation of the trajectories obtained at the anti-integrable limit should be very fruitful for finding new results.

4.1 New theorems on the continuation of anti-integrable trajectories and periodic cycles. At the anti-integrable limit, the periodic cycles are simply obtained by choosing a coding sequence which fulfills condition (14-a) for all i. For a given periodic cycle $\mathbb{C}$ with period s of the standard map $\mathcal{S}(3)$, the symplectic matrix $\overline{\overline{P}}$ of the linearized map of $\mathcal{S}^s$, is the ordered product

(23-a) $$\overline{\overline{P}} = \prod_{i=s}^{1} \overline{\overline{P}}_i \qquad \text{with}$$

(23-b) $$\overline{\overline{P}}_i = \begin{pmatrix} 1 + k\cos u_i & 1 \\ k\cos u_i & 1 \end{pmatrix}$$

This matrix $\overline{\overline{P}}$ has two eigenvalues λ_1, λ_2 (with product 1) which do not depend on the initial point of the periodic cycle. $\mathbb{C}$ is called hyperbolic without reflection when λ_1, λ_2 are both real positive. It is hyperbolic with reflection when both are real negative. It is elliptic when both are complex conjugate with modulus 1. Marginal cases with degenerate eigenvalues are called parabolic without or with reflection whether the sign of $\lambda_1 = \lambda_2$ is positive or negative respectively. Hyperbolic cycles without reflection have a Poincaré index -1, both hyperbolic cycles with reflection and elliptic cycles have a Poincaré index $+1$.

For a s-periodic cycle, the eigenvectors of the QEAM $\overline{\overline{M}}$ obey the Bloch-Floquet theorem which states that the eigenvectors of $\overline{\overline{M}}$ have the form $\overline{\varepsilon}_q = \exp(iqn)\chi_n\}$ where q is some arbitrary number and the sequence $\{\chi_n\}$ is s-periodic: that is $\chi_{n+s} = \chi_n$ for all n. All these eigenstates are uniformly bounded and thus, zero belongs to the spectrum of the QEAM (17) iff (if and only if) the linear set of equations

(24-a) $$\overline{\overline{M}}\overline{\varepsilon} = 0$$

has a bounded solution $\overline{\varepsilon} = \{\varepsilon_i\}$. Using the definition (17), it is straightforward to find that we have for this solution

(24-b) $$\begin{pmatrix} \varepsilon_{s+1} \\ \varepsilon_{s+1} - \varepsilon_s \end{pmatrix} = \overline{\overline{P}} \begin{pmatrix} \varepsilon_1 \\ \varepsilon_1 - \varepsilon_0 \end{pmatrix}$$

This condition is fulfilled iff the eigenvalues λ_1 and λ_2 of $\overline{\overline{P}}$ have both modulus 1 in order that sequence $\{\varepsilon_i\}$ does not diverge neither for $i \to +\infty$ nor for $i \to -\infty$. Thus, the gap parameter of a periodic cycle is zero iff

(24-c) $$-2 \leq \text{ Trace } \overline{\overline{P}} \leq 2$$

Consequently, the hyperbolic cycles have strictly non-zero gap parameter, while the elliptic and parabolic periodic cycles have zero gap parameter. We now prove a theorem which states that the parity of the number of balls in a period of the periodic cycle, on top of the barrier of the periodic potential $V(x)$, determines at large k whether the Poincaré index of the periodic cycle is $+1$ or -1.

THEOREM 4. *For k fulfilling (8-b), a periodic cycle determined by a coding sequence fulfilling (14-a) is hyperbolic without reflection (Poincaré index-1) when the number n_s of odd m_i for $1 \leq i \leq s$ is even, and it is hyperbolic with reflection (Poincaré index +1) when n_s is odd.*

Proof. Let us consider a s-periodic cycle C characterized by a coding sequence $\{m_i\}$ fulfilling (14-a). For $k \to +\infty$, condition (9-b) and definition (24-b) readily imply

(25-a) $$\text{Trace } \bar{\bar{P}} \approx (-1)^{M_s} \; k^s$$

with

(25-b) $$M_s = \sum_{i=1}^{s} m_i = 2 \sum_{i=1}^{s} \text{Int}\left(\frac{m_i}{2}\right) + n_s$$

For large enough k, theorem 2 asserts that these periodic cycles have a finite gap parameter and thus are hyperbolic. The sign of $(-1)^{M_s} = (-1)^{n_s}$, which is also the sign of the eigenvalues of $\bar{\bar{P}}$ determines whether these hyperbolic cycles are without reflection (n_s even) or with reflection (n_s odd). When k decreases, since these periodic cycles $\{u_i\}$ continuously depend on k, Trace $\bar{\bar{P}}$ is also a continuous function of k. Thus, the sign of Trace $\bar{\bar{P}}$, cannot change without fulfilling (24-c) over some interval in k. If this situation occurs, the gap parameter of the periodic cycle C vanishes. But according to theorem 1, k must be smaller than $K(b)$, the value given by (8-b), since according to theorem 2 its gap parameter remains strictly positive for $k > K(b)$.

□

We present now a theorem which asserts that the trajectories of the standard map which are "too much stressed", cannot be continued to small values of k. For a coding sequence $\{m_i\}$, we define

(26-a) $$b_i = m_{i+1} + m_{i-1} - 2m_i$$

and from now on, we choose systematically for b in (8-a)

(26-b) $$b = \operatorname*{Sup}_i |b_i| \; < +\infty$$

This integer b is called the stress parameter of the trajectory. Theorem 1 states that this trajectory can be continued from $k = +\infty$ to $k = \sqrt{16 + (b+2)^2\pi^2}$. Theorem 3 states that it can be uniquely continued till the gap parameter vanishes. The next theorem in some sense is a reciprocal theorem, because it gives a lower bound in k beyond which this uniform continuation is impossible.

THEOREM 5. *Let us consider either an arbitrary coding sequence* $\{m_i\}$ *with* $b \geq 11$, *or an arbitrary coding sequence with* m_i *even for all* i, *with* $b \geq 9$ *(*b *is defined by (26)). Then the anti-integrable trajectory* $\{u_i(k)\}$ *of the standard map, which is obtained from theorem 1 with this coding sequence* $\{m_i\}$ *cannot be continued uniformly for* $k < (b-4)\pi$.

This theorem applies for any possible uniform continuation of an anti-integrable trajectory periodic or not, even when this continuation becomes non unique and when the gap parameter becomes zero. This situation occurs for example at bifurcations where several paths could be chosen, or for elliptic periodic cycles which could be continued by KAM tori. This theorem proves that most trajectories which exist at the anti-integrable limit must disappear when k decreases below some critical value which depends on the trajectory.

Proof. Let us consider a trajectory $\{u_i(k)\}$ of the standard map which is obtained from theorem 1 for $k \geq K(b)$. We have $\{u_i(\infty)\} = \{m_i\pi\}$. According to the theorems 2 and 3, it can be continued as a uniformly continuous function of k down to some values of k. Our purpose is to make a lower bound in k below which any uniform continuation of $\{u_i(k)\}$ is strictly impossible. We set

(27-a) $$u_i(k) = m_i\pi + x_i(k)$$

Then, the stationarity equation (2-b) becomes

(27-b) $$k(-1)^{m_i} \sin\ x_i = b_i\pi + x_{i+1} + x_{i-1} - 2x_i$$

We define

(28-a) $$x(k) = \operatorname*{Sup}_i |x_i(k)|$$

which is a continuous function of k. Eq.(27-b) readily implies

(28-b) $$k \geq |b_i|\pi - 4x(k)$$

Since (28-b) is true for all i, we have

(28-c) $$4x(k) \geq b\pi - k\ .$$

Consequently, when $k < (b-4)\pi$, we should have $x(k) > \pi$. Since $x(k)$ is a continuous function of k and $\ll_{k\to+\infty} x(k) = 0$, there exists k_0 fulfilling

(29-a) $$(b-4)\pi \leq k_0 < K(b) = \sqrt{16 + (b+2)^2\pi^2}$$

and such that

(29-b) $$x(k_0) = \pi \qquad \text{and}$$

(29-c) $$x(k) < \pi \qquad \text{for}\ \ k > k_0.$$

For proving theorem 5, we prove that it is impossible to fulfill (29) when $b \geq 11$, or when $b \geq 9$ if all m_i are even.

Eq.(27-b) can be written either as $f_1(x_i) = b_i\pi + x_{i+1} + x_{i-1}$ (m_i even) or as $f_2(x_i) = b_i\pi + x_{i+1} + x_{i-1}$ (m_1 odd) with $f_1(x) = 2x + k\sin x$ and $f_2(x) = 2x - k\sin x$. When $k > 2$, function $f_1(x)$ has a maximum and a minimum on the interval $[-\pi, \pi]$, at a_1 and $-a_1$ respectively, with $a_1 = \frac{\pi}{2} + \text{Arcsin}\,\frac{2}{k}$ (see fig.2). It is readily found that $|f_1(\pm a_1)| > A_1(k) = k + \pi > 0$. When $k > 2$, function $f_2(x)$ also has two extrema on the interval $[-\pi, \pi]$ at $\pm a_2$, with $a_2 = \frac{\pi}{2} - \text{Arcsin}\,\frac{2}{k}$ and we have $|f_2(\pm a_2)| > A_2(k) = k - \pi$.

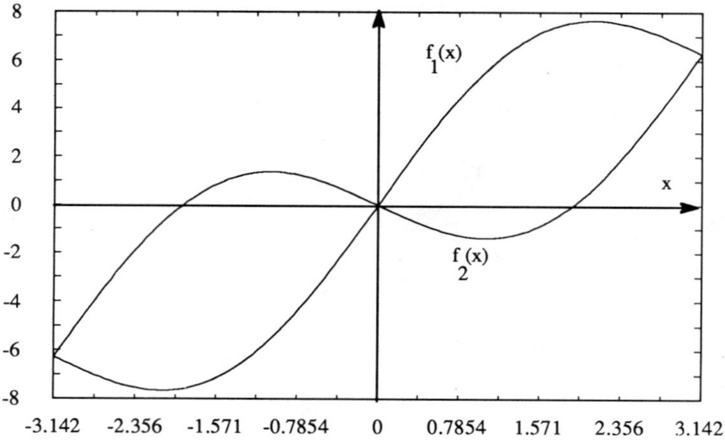

FIGURE 2: *Graph of functions $f_1(x)$ and $f_2(x)$ on the interval $[-\pi, +\pi]$ for $k = 4$.*

If for k in the interval $[k_0, +\infty]$, $|b_i\pi + x_{i+1} + x_{i-1}|$ is always strictly smaller than $|f_2(\pm a_2(k))| < |f_1(\pm a_1(k))|$ and since $x_i(k)$ is a continuous function of k fulfilling $\ll_{k\to+\infty} x_i(s) = 0$, we necessarily have for all $k \geq k_0$, $|x_i(k)| < a_1(k)$ when m_i is even, or $|x_i(k)| < a_2(k)$ when m_i is odd. Since $|b_i\pi + x_{i+1} + x_{i-1}| < (|b_i| + 2)\pi$, this condition is a fortiori fulfilled when

(30-a) $$(|b_i| + 2)\pi \leq (b - 5)\pi$$

since we have

(30-b)

$$(|b_i| + 2)\pi \leq (b - 5)\pi \leq A_2(k_0) = k_0 - \pi \leq A_2(k) < |f_2(\pm a_2(k))| < |f_1(\pm a_1(k))|$$

Consequently, for m_i even or odd, we necessarily have

(31-a) $$|x_i(k_0)| < a_1(k_0) \neq \pi \qquad \text{when}$$

(31-b) $$|b_i| \leq b - 7$$

Since for $k = k_0$, we have by (29), $\operatorname{Sup}_i |x_i(k_0)| = x(k_0) = \pi$, there exists a subsequence $\{x_{i_\nu}\}$ of $\{x_i\}$ such that $\ll_{\nu\to\infty} |x_{i_\nu}| = \pi$. We can choose this subsequence such that $\pi \geq |x_{i_\nu}| > a_1(k_0)$, so that inequalities (31) imply

(32-a) $$|b_{i_\nu}| > b - 7 \qquad \text{or} \qquad |b_{i_\nu}| \geq b - 6 = b - 7 + 1$$

for all ν. Then, eq.27-b yields

(32-b) $$\lim_{\nu\to\infty} b_{i_\nu}\pi = \ll_{\nu\to\infty} (x_{i_\nu+1} + x_{i_\nu-1} - 2x_{i_\nu})$$

which implies because of (32-a)

(32-c) $$(b-6)\pi \leq \ll_{\nu\to\infty} |b_{i_\nu}|\pi \leq 4\pi$$

This condition implies $b \leq 9$ which is in contradiction with the hypothesis of theorem 5. Consequently, $\{u_k(k)\}$ cannot be continued uniformly down to k_0. If we assume that the coding sequence only contains even m_i, the reader can easily check that the condition $b \geq 9$ is sufficient for proving the theorem since $A_1 = k_0 + \pi$ replaces $A_2 = k_0 - \pi$ in (30).

□

A straightforward consequence of theorem 5, is the existence of infinitely many elliptic periodic cycles in the standard map, for k arbitrarily large. We have:

Corollary of Theorem 5. *Let B be an arbitrary positive number and $s \geq 2$ an arbitrary integer,*

then the standard map exhibits infinitely many elliptic periodic cycles with period s, for k in finite intervals $]k_1, k_2[$ with $b < k_1 < k_2$

(of course, these intervals are generally different for each elliptic periodic cycle)

Proof. For $s \geq 2$ and arbitrary integer r (with $\frac{r}{s}$ irreducible), let us consider any coding sequence $\{m_i\}$ fulfilling (14-a) with $b \geq \operatorname{Max}\left(11, 4 + \frac{B}{\pi}\right)$ and such that the number n_s of odd m_i (for $1 \leq i \leq s$) be an odd number. According to theorem 4, the periodic cycle $\{u_i\}$ which exists for $k > K(b)$ is hyperbolic with reflection.

This periodic cycle $\{u_i\}$ corresponds to an extrema of the finite truncated variational form (1)

(33-a) $$\Phi_s(\{u_i\}) = \sum_{i=1}^{S} [W(u_{i+1} - u_i) + V(u_i)]$$

with the boundary condition

(33-b) $$u_{s+1} = u_1 + 2r\pi$$

This form only depends on s variables $u_i (i = 1, 2, \dots s)$. As for the infinite case (16-a), we consider the finite QEAM $\overline{\overline{M}}_s$ which is the $s \times s$ matrix of second order derivatives of $\Phi_s(\{u_i\})$ at the extrema $\{u_i\}$. It is easy to prove recursively the identity valid for $s \geq 2$

(34-a) $$2 + \frac{1}{C^s} \operatorname{Det}(\overline{\overline{M}}_s) = \operatorname{Trace} \overline{\overline{P}}$$

where $\operatorname{Det}(\overline{\overline{A}})$ denotes the determinant of matrix $\overline{\overline{A}}$. $\overline{\overline{P}}$ is defined by (23) for $\{u_i(k)\}$. Since this extrema $\{u_i(k)\}$ of $\Phi_s(\{u_i\})$ cannot be continued for $k < (b-4)\pi$, there exists $k_1 > (b-4)\pi > B$ such that $\operatorname{Det}(\overline{\overline{M}}_s(k_1)) = 0$. Then, we have

(34-b) $$\operatorname{Trace} \overline{\overline{P}}(k_1) = 2$$

But since for $k > K(b)$ (theorem 4) this periodic cycle is hyperbolic with reflection, Trace $\overline{\overline{P}}(k)$ is negative. Since Trace $\overline{\overline{P}}(k)$ depends continuously on k, there exists a finite open interval $]k_1, k_2[$ such that

(35-a) $$-2 < \operatorname{Trace} \overline{\overline{P}}(k) < 2$$

for

(35-b) $$B < k_1 < k < k_2 \ .$$

Then the s-periodic cycle $\{u_i\}$ is elliptic, which proves the corollary.

□

Note that k_1 may be different of k_0 if this elliptic s-periodic cycle undergoes a bifurcation for $k = k_1$. It is also possible that at $k = k_0 = k_1$, this elliptic s-periodic cycle annihilates with another s-periodic cycle which must be hyperbolic without reflection in order to fulfill the Poincaré index rule.

Remark on the Bifurcations of Periodic Cycles. The QEAM $\overline{\overline{M}}_s$ of a periodic cycle $\{u_k(k)\}$ can be written

(36-a) $$\overline{\overline{M}}_s = \overline{\overline{D}}_s + \overline{\overline{F}}_s$$

where $\overline{\overline{D}}_s = \{D_{i,j}\}$ with $D_{i,j} = (2 + k \cos u_i(k))\delta_{i,j}$ is the diagonal part of $\overline{\overline{M}}_s$ and $\overline{\overline{F}}_s$ is its off-diagonal part with Hermitian norm $\|\overline{\overline{F}}_s\| = 2$. We set $\overline{\overline{M}}_s(x) = \overline{\overline{D}}_s + x\overline{\overline{F}}_s$. The eigenvalues $\lambda_\nu(x)$ of $\overline{\overline{M}}_s(x)$ are continuous functions of x and for $x \neq 0$, are always non degenerate. For $x = 0$, we have $\lambda_{\nu_i}(0) = (2 + k \cos u_i)$. Using standard perturbation theory, it comes out $|\frac{d\lambda_\nu(x)}{dx}| \leq \|\overline{\overline{F}}_s\| = 2$ and consequently, the set of s eigenvalues $\{\lambda_\nu\}$ of $\overline{\overline{M}}_s$ can be ordered as $\{\lambda_{\nu_i}\}$ in such a way that we have for $1 \leq i \leq s$

$$|\lambda_{\nu_i} - (2 + k\cos u_i)| \leq 2 \ . \tag{36-b}$$

Let us assume that for given i, $2 + k\cos\ u_i(k)$ crosses the interval $[-2, 2]$ for k in the interval $[k_1, k_2]$ (that is there exists $k_1' < k_1 < k_2 < k_2'$ such that

$$-2 \leq 2 + k\cos\ u_i(k) \leq 2 \qquad \text{for } k_1 \leq k \leq k_2 \qquad \text{and} \tag{37-a}$$

either (crossing in positive direction)

$$2 + k\cos\ u_i(k) < -2 \qquad \text{for } k_1' < k < k_1 \qquad \text{and} \tag{37-b}$$

$$2 + k\cos\ u_i(k) > 2 \qquad \text{for } k_2 < k < k_2' \tag{37-c}$$

or (crossing in negative direction)

$$2 + k\cos\ u_i(k) > 2 \qquad \text{for} \qquad k_1' < k < k_1 \qquad \text{and} \tag{37-d}$$

$$2 + k\cos\ u_i(k) < -2 \qquad \text{for} \quad k_2 < k < k_2' \quad), \tag{37-e}$$

then an eigenvalue of $\overline{\overline{M}}_s$ may change its sign. If a single eigenvalue λ_{ν_i} of $\overline{\overline{M}}_s$ changes its sign for $k = k_0$ with $k_1 \leq k_0 \leq k_2$, then $\det(\overline{\overline{M}}_s) = 0$ and the corresponding periodic cycle must undergo a bifurcation for $k = k_0$.

However, for asserting that an eigenvalue of $\overline{\overline{M}}_s$ does change its sign, we have to be sure that for other values of i, no crossing occurs in the opposite direction, in an interval $[k_1^*, k_2^*]$ which overlaps with $[k_1, k_2]$. It is in principle possible to have this situation without vanishing any eigenvalues for $\overline{\overline{M}}_s$. However, the coding sequences $\{m_i\}$ can be chosen in order that there is no ambiguity.

For example, let us consider a coding sequence fulfilling (14-b) with all m_i even. According to theorem 4, for large k all the eigenvalues of $\overline{\overline{M}}_s$ are positive. When $x(k) = \operatorname{Sup}_i |u_i(k) - m_i\pi| = \frac{\pi}{2} + \operatorname{Arcsin}\frac{4}{k}$, one eigenvalue (at least) of $\overline{\overline{M}}_s$ is negative or zero. Then it comes out that this s-periodic cycle which is hyperbolic without reflection (index -1) for large k, must undergo a bifurcation when decreasing k to $k = k_b > (b-4)\pi$. (This loose bound is easily obtained from inequality (28-c)). According to standard bifurcation theory (Poincaré index sum rule), this bifurcation must involve at least an elliptic s-periodic cycle (with index $+1$). Note that this result yields another proof for the corollary of theorem 5. Similar result can be obtained with all m_i odd. However, this remark which only applies to certain coding sequences, is different from those predicted by theorem 5, since the existence of bifurcations does not prevent the continuation of a periodic cycle as a function of k.

For completing our study on trajectory bifurcations, let us now analyze some examples of bifurcation and anti-integrable trajectories.

4.2 Example of periodic cycles bifurcations. Our purpose in this subsection is to show on some examples, how periodic cycles which are obtained at large k from given coding sequences can be followed continuously to small values of k. The simplest examples are the 1-periodic cycles of the standard map (fixed points) which are simply given by the extrema of $V(x)$. Their coding sequences have the form $m_i = ip + q$ where p and q are arbitrary integers. Their stress parameter b defined by (26) is always zero. Let us consider for example, the fixed point C_0 obtained by uniform continuation of the 1-periodic cycle generated at large k by the coding sequence $m_i = 1$ for all i. The trajectory corresponding to C_0 is described by $u_i = \pi$ for all i and all k. This 1-periodic cycle is hyperbolic with reflection for $k > 4$, and becomes elliptic for $k < 4$.

Let us consider now the less trivial example which is the 2-periodic cycle called C_1 generated by the coding sequence

$$\text{(38-a)} \qquad m_i = \frac{1+(-1)^i}{2} \quad , \ \{m_i\} = \{\ldots, 1, 0, 1, 0, 1, 0 \ldots\}$$

It fulfills condition (8-a) with stress parameter $b = 2$. Theorem 1 states that for $k > 4\sqrt{\pi^2+1}$, there exists a unique trajectory $\{u_i\}$ of the standard map such that $|u_i - m_i\pi| < \frac{\pi}{2}$. The corresponding 2-periodic cycle can be explicitly calculated

$$\text{(38-b)} \qquad u_i = \alpha + (-1)^i \frac{\pi}{2} \qquad \text{where } 0 < \alpha \leq \frac{\pi}{2} \text{ is given by the equation}$$

$$\text{(38-c)} \qquad 2\pi = k \cos \alpha \; .$$

This explicit formula shows that this 2-periodic cycle can be continued analytically in the interval $2\pi \leq k < 4\sqrt{\pi^2+1}$. The gap parameter of this periodic cycle is found to be non vanishing only for $2\sqrt{\pi^2+1} < k$ while this 2-periodic cycle is hyperbolic with reflection. The gap parameter of C_1, is found to be zero when $2\pi < k < 2\sqrt{\pi^2+1}$ and then C_1 is elliptic. Theorem 3 does not apply. Indeed, according to the KAM theory for elliptic fixed points, in any neighborhood of this 2-periodic cycle, there exist trajectories dense on a couple of KAM tori ("islands"). Consequently, for $2\pi < k < 2\sqrt{\pi^2+1}$, the k-continuation of the trajectory corresponding to the 2-periodic cycle is not unique. In the *full space* of trajectories of the standard map, we could consider each value of k where this periodic cycle is elliptic as a *generalized* bifurcation point.

In fact, for insuring a unique well-defined uniform continuation, one (implicitly) requires that this trajectory *remains* a 2-periodic cycle. Then the periodic cycle remains described by the explicit form (38-b). In the *restricted* subspace of 2-periodic cycles, the *standard* bifurcation of C_1 only occurs at $k = 2\pi$ when eq.(38-c) has no real solution. Then the elliptic 2-periodic cycle C_1 together with another elliptic periodic cycle C_1' and an hyperbolic 2-periodic cycle C_2 merge into an elliptic periodic cycle. The 2-periodic cycle C_1' is obtained from C_1 by the symmetry $u_i \to 2\pi - u_i$. Thus, C_1' has the same properties as C_1. According to theorem 1, it is obtained at large k from the coding sequence

(38-d) $$m_i = \frac{3+(-1)^i}{2} \quad , \ \{m_i\} = \{\ldots 2,1,2,1,2,1\ldots\} \qquad \text{with} \quad b = 2\ .$$

The hyperbolic cycle C_2 is described at large k by the coding sequence

(39-a) $$m_i = 1 + (-1)^i \qquad \{m_i\} = \{\ldots, 2,0,2,0,2,0,\ldots\} \quad \text{with} \quad b = 4$$

and is explicitly given by

(39-b) $$u_i = \pi + (-1)^i \beta \qquad \text{where } 0 \le \beta < \pi \text{ is given by the equation}$$

(39-c) $$k \sin\beta = 4\beta$$

This 2-periodic cycle is hyperbolic without reflection for $k > 2\pi$. It becomes elliptic for $4 < k < 2\pi$. For $k < 4$, this 2-periodic elliptic cycle merges with the 1-periodic cycle C_0. Note that C_2 (and also C_1 and C_1'), can be continued uniformly to $k = 0$ via the periodic cycle C_0. Thus, theorem 5 cannot be extended to trajectories with values of b smaller or equal to 4.

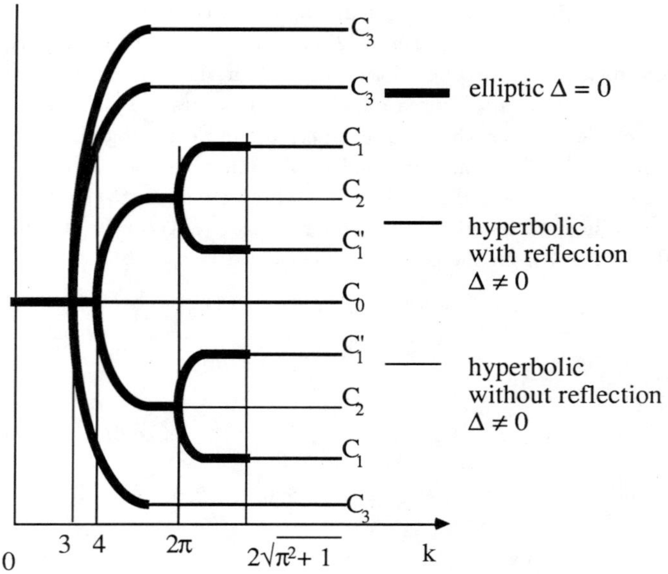

FIGURE 3: *Bifurcation scheme of the 3-periodic cycle C_3, of the 2-periodic cycles C_1, C_1', C_2 and of the 1-periodic cycle C_0 described in the text. All these periodic cycles come from trajectories which are given by theorem 1 at large k and thus can be associated with a coding sequence $\{m_i\}$.*

The 3-periodic coding sequence

(40-a) $$m_i = i \text{ modulo } 3 \quad \{m_i\} = \{\ldots, 0, 1, 2, 0, 1, 2, 0, 1, 2 \ldots\}$$

with stress parameter $b = 3$, corresponds to the 3-periodic cycle C_3 :

(40-b) $$u_i = \pi + (m_i - 1)\alpha \quad , \text{ where } 0 < \alpha < \pi \text{ is the solution of}$$

(40-c) $$k \sin \alpha = 3\alpha \ ,$$

which exists only for $k > 3$. This 3-periodic cycle is hyperbolic with reflection (with a non vanishing gap parameter) for $k > k_0 \approx 4.096$, and becomes elliptic for $k < k_0$. At $k = 3$, this 3-periodic cycle merges with the elliptic 1-periodic cycle C_0. In fact, there exist infinitely many other periodic cycles which merge with C_0. The bifurcation scheme of the periodic cycles discussed in this subsection is shown in fig.3.

As suggested by theorem 5, some of the periodic cycles generated at large k have a very simple bifurcation: they just annihilate in pairs when decreasing k. Take for example the 3-periodic cycle C_3' generated by the coding sequence

(41-a)

$$m_i = 2\left(\mathrm{Int}\left(\frac{i}{3}\right) - \mathrm{Int}\left(\frac{i-1}{3}\right)\right) \qquad \{m_i\} = \{\ldots, 0,0,2,0,0,2,0,0,2\ldots\}$$

with stress parameter $b = 4$, and the 3-periodic cycle C_3'' generated by

(41-b) $$m_i = \left(\mathrm{Int}\left(\frac{i}{3}\right) - \mathrm{Int}\left(\frac{i-1}{3}\right)\right) \qquad \{m_i\} = \{\ldots, 0,0,1,0,0,1,0,0,1\ldots\}$$

with $b = 1$. A tedious calculation allows one to show that C_3' and C_3'' are both defined for $k > \frac{8\pi}{3}$ and annihilate with each other at $k = \frac{8\pi}{3}$. (At the bifurcation point, the degenerate 3-periodic cycle corresponds to the trajectory $\{u_i\} = \left\{\ldots \frac{\pi}{6}, \frac{\pi}{6}, \frac{3\pi}{2}, \frac{\pi}{6}, \frac{\pi}{6}, \frac{3\pi}{2} \ldots\right\}$). For $k > \frac{8\pi}{3}$, the 3-periodic cycle C_3 is always hyperbolic without reflection with a non vanishing gap parameter, while for large k the 3-periodic cycle C_3' is hyperbolic with reflection with a non vanishing gap parameter. At a certain value of $k > \frac{8\pi}{3}$, it becomes elliptic with a zero gap parameter.

Although in principle these bifurcations can be reversed, we never observed examples of inverted bifurcation, that is pairs of periodic cycles annihilating each other when *increasing* k (instead of decreasing k).

There also exist periodic cycles which never bifurcate. Take for example, the 2-periodic cycle which corresponds to the coding sequence $\{m_i\}$ chosen as

(42-a) $$m_i = 2\,\mathrm{Int}\left(\frac{i}{2}\right) \qquad \{m_i\} = \{\ldots 0,0,2,2,4,4,6,6\ldots\}$$

with $b = 2$. In fact, this 2-periodic cycle just represents a ground-state with commensurability 2 ($l = \pi$); it corresponds to the trajectory $\{u_i\}$ with

(42-b) $$u_i = m_i\pi + (-1)^{i+1}\alpha$$ where α is the solution of the equation

(42-c) $$k \sin\alpha + 4\alpha = 2\pi\ ,$$

which exists for any postive k and is unique in the interval $\left[0, \frac{\pi}{2}\right]$.

More generally, the numerical studies on the FK model suggested that the ground-states $\{u_i(k)\}$ are uniformly continuous functions of k. (However, note that this property has been found to be wrong for other twist maps [9]). As it will be shown in the next section, the trajectory $\{u_i(k)\}$ associated with the coding sequence $m_i = 2\,\mathrm{Int}\left(\frac{il+\alpha}{2\pi}\right)$ where $\zeta = \frac{l}{2\pi}$ is a given irrational number and α is an arbitrary phase, corresponds to the ground-state with irrational incommensurability ratio ζ (for large enough k). When k decreases from infinity to a certain critical value $k_c(\zeta)$, the gap parameter of $\{u_k(k)\}$ has been numerically found to be strictly positive. It has been found to be zero for $0 \leq k \leq k_c(\zeta)$. According to the KAM theory, $k_c(\zeta)$ is generally not zero.

In conclusion, these examples suggest that many trajectories of the standard map (maybe all of them) can be obtained by uniform continuation of trajectories characterized by a coding sequence, which are predicted to exist at the anti-integrable limit (theorem 1). An important question is:

Could a coding sequence be found for any trajectory of the standard map?

Theorem 5 shows that for most coding sequences, the continuation cannot be done at small value of k. It also suggest that the entropy of the standard map decreases when k decreases, by "pruning" the set of coding sequences into a smaller and smaller subset. Let us also mention now the two conjectures given in ref.6.

Conjectures: *The Lebesgue measure on the cylinder of the set of trajectories of the standard map which have a non-vanishing gap parameter Δ, is zero.*

The closure of this set has a non-vanishing Lebesgue measure and perhaps is the full space.

Following the trajectories as uniform continuous functions of k till the gap parameter vanishes (generalized bifurcation), could give information on the property of the full chaotic region of the standard map and on its measure.

5. Minimal orbits close to the anti-integrable limit. In this last section, we essentially focus on the minimal orbits of the standard map. They are associated with the minimum energy configurations of the FK model. By definition, a minimum energy configuration $\{u_i\}$ of the FK model has the property that any configuration $\{u_i + \delta_i\}$ where $\{\delta_i\}$ is arbitrary with a compact support, has a larger "energy" than the initial configuration $\{u_i\}$. ($\{\delta_i\}$ is said to have a compact support when there exists a finite number N, such that $\delta_i = 0$ for $|i| \geq N$).

In other words, we have for any N and any $\{\delta_i\} \neq 0$ with $\delta_i = 0$ for $|i| \geq N$:

$$\Delta\Phi(\{u_i\}) = \sum_{i=-N}^{i=N} \left(L(u_{i+1} + \delta_{i+1}, u_i + \delta_i) - L(u_{i+1}, u_i)\right) > 0 \tag{43}$$

with $L(x, y) = W(x, y) + V(x)$ defined by (1-b) and (1-c). The minimum energy configurations $\{u_i\}$ are minima of the variational form (1-a) and thus fulfill obviously the extremalization equation (2-b). They correspond to a subset of trajectories of the standard map which is called $\mathcal{Q}$.

The ground-states are defined as the minimum energy configurations which correspond to recurrent trajectories. This set is called $\mathcal{G}$ and we have $\mathcal{Q} \supseteq \mathcal{G}$.

The properties of these configurations (or trajectories of the standard map) are described in refs.10 and 12. Similar results were found by J. Mather [11]. (A detailed mathematical analysis can be found in ref.7).

Let us briefly recall the main results.

1- For any minimum energy configuration, the limit

$$l = \ll_{N'-N\to\infty} \frac{u_{N'} - u_N}{N' - N} \tag{44}$$

is defined, and reciprocally for any l there exists a set $\mathcal{Q}_l$ of minimum energy configurations fulfilling (44).

2- Incommensurate Ground-states. For $\zeta = \dfrac{l}{2\pi}$ irrational, the set $\mathcal{Q}_l$ is well ordered which means that for two configurations $\{u_i\}$ and $\{v_i\}$ in $\mathcal{Q}_l$ with $\{u_i\} \neq \{v_i\}$, we have for all i, either $u_i < v_i$ or $u_i > v_i$

The whole set $\mathcal{G}_l$ of ground-states $\{u_i\}$ with $\mathcal{Q}_l \supseteq \mathcal{G}_l$, is described by a single "hull function" $f(x)$ as

$$u_i = f(il + \alpha) \tag{45-a}$$

where the phase α is arbitrary. The hull function $f(x)$ which depends on l, has the following properties:

a) $f(x)$ is monotonous strictly increasing

b) $f(x) - x = g(x)$ is 2π periodic

c) $f(x)$ is either right continous or left continuous or is continuous. When it is discontinuous,

both the left continuous and the right continuous determination, define ground-states by (44-a).

Then, the corresponding trajectory in the standard map $\{u_i, p_i\}$ is dense on an invariant circle when $f(x)$ is continuous or on a Cantorus when $f(x)$ is discontinuous (Aubry Mather set).

3- Commensurate Ground-states. For $\zeta = \dfrac{l}{2\pi} = \dfrac{r}{s}$ rational (r and s are relatively prime integers), the corresponding ground-states $\{u_i\}$ fulfill

$$u_{i+s} = u_i + 2\pi\ r \tag{45-b}$$

and thus are represented by s-periodic cycles in the standard map. The set of commensurate ground-states $\mathcal{G}_l$ is closed and well ordered.

4- Discommensurations. For $\zeta = \dfrac{l}{2\pi} = \dfrac{r}{s}$ rational, unless the corresponding ground states $\{u_i\}$ form a continuum of configurations $\{u_i(\alpha)\}$ with $\alpha \in \mathbb{R}$, we have $\mathcal{Q}_l \neq \mathcal{G}_l$. $\mathcal{Q}_l$ contains minimum energy configurations called discommensurations. For each discommensuration $\{v_i\}$, there exists two distinct (consecutive) commensurate ground-states $\{u_i^+\}$ and $\{u_i^-\}$ in $\mathcal{G}_l$ with $\{u_i^+\} > \{u_i^-\}$ and such that $\{u_i^+\} > \{v_i\} > \{u_i^-\}$. For the advanced discommensurations, we have

$$\text{(46-a)} \qquad \lim_{i\to+\infty} (u_i^+ - v_i) = 0 \quad , \quad \lim_{i\to-\infty} (v_i - u_i^-) = 0$$

and the property

$$\text{(46-b)} \qquad u_{i+s} - u_i - 2\pi\ r > 0 \qquad \text{for all } \ i$$

For the delayed discommensurations, we have

$$\text{(46-c)} \qquad \ll_{i\to+\infty} (v_i - u_i^-) = 0 \qquad \text{and} \qquad \ll_{i\to-\infty} (u_i^+ - v_i) = 0$$

$$\text{(46-d)} \qquad u_{i+s} - u_i - 2\pi\ r < 0 \qquad \text{for all } \ i$$

Close enough to the anti-integrable limit, we can determine the coding sequences of the whole set of minimum energy configurations. We have:

THEOREM 6. *For $k \geq K(2) = 4\sqrt{1+\pi^2}$, the whole set of minimum energy configurations of the FK model belongs to the set of stationary configurations obtained by theorem 1 from coding sequences with stress parameter $b = 2$ (for $l \neq 2p\ \pi$ with p integer).*

The coding sequences of the ground-states of $\mathcal{G}_l$ with atomic mean distance l (44) which is either commensurate or incommensurate, are given by

$$\text{(47-a)} \qquad m_i = 2\,\mathrm{Int}\left(\frac{il+\alpha}{2\pi}\right)$$

where α is an arbitrary phase.

The coding sequences of the advanced discommensurations for $\zeta = \dfrac{l}{2\pi} = \dfrac{r}{s}$ rational, are given by

$$\text{(47-b)} \qquad m_i = 2\,\mathrm{Int}\left(\frac{ir}{s} + \alpha_i\right)$$

where $\dfrac{p-1}{s} < \alpha_i < \dfrac{p}{s}$ for $i < n$ and $\dfrac{p}{s} < \alpha_i < \dfrac{p+1}{s}$ for $n \leq i$, and the phase index p and the location n are arbitrary integers.

The coding sequences of the delayed (or retarded) discomensurations are also given by (47-b) where $\dfrac{p}{s} < \alpha_i < \dfrac{p+1}{s}$ for $i < n$, and $\dfrac{p-1}{s} < \alpha_i < \dfrac{p}{s}$ for $n \leq i$.

Proof. For any minimum energy configuration, it readily comes out from the definition that u_i is the value of x which yields the absolute minimum to the local potential $L(u_{i+1}, x) + L(x, u_{i-1})$ or equivalently for the FK model, of function

$$\text{(48-a)} \qquad U(x) = \lambda(1 - \cos\ x) + C\ x^2 - C(u_{i+1} + u_{i-1})x$$

The absolute minima u_i of $U(x)$, fulfills

(48-b) $$|u_{i+1} + u_{i-1} - 2u_i| < 2\pi \ .$$

Thus, eq.2-b yields $k|\sin u_i| < 2\pi$. For $k > 2\pi$, u_i must belong to the union of open intervals:

(49-a) $$u_i \in \bigcup_{n\in\mathbb{Z}}]-x_0 + n\pi,\ x_0 + n\pi[= I_0$$

where

(49-b) $$0 \leq x_0 = \operatorname{Arcsin}\left(\frac{2\pi}{k}\right) \leq \frac{\pi}{2} \ .$$

Otherwise, because of its definition, the QEAM of a minimum energy configuration $\{u_i\}$ must be strictly positive. A necessary condition is that the diagonal terms of $\overline{\overline{M}}(\{u_i\})$ defined by (16-a) or (17) be strictly positive

(50-a) $$2 + k\cos\ u_i > 0 \ ,$$

which implies

(50-b) $$u_i \in \bigcup_{n\in\mathbb{Z}}]-x_1 + 2n\pi,\ x_1 + 2n\pi[= I_1$$

where

(50-c) $$\frac{\pi}{2} < x_1 = \operatorname{Arccos}\left(-\frac{2}{k}\right) = \frac{\pi}{2} + \operatorname{Arcsin}\frac{2}{k}$$

When $x_1 \leq \pi - x_0$, or equivalently when

(51-a) $$k \geq 2\sqrt{1+\pi^2}$$

we have

(51-b) $$u_i \in I_0 \cap I_1 = \bigcup_{n\in\mathbb{Z}}]-x_0 + 2n\pi,\ x_0 + 2n\pi[\ .$$

We define

(52-a) $$m_i = 2\,\operatorname{Int}\left(\frac{u_i + \frac{\pi}{2}}{2\pi}\right)$$

so that when (51-a) is fulfilled, (51-b) implies $|u_i - m_i\pi| < x_0 < \frac{\pi}{2}$ and

(52-b) $$\{u_i\} \in E\left(\{m_i\}; \frac{\pi}{2}\right) \ .$$

When, $k > K(2) = 4\sqrt{1+\pi^2}$, theorem 1 asserts that there exists one and only one stationary configuration in $E\left(\{m_i\};\frac{\pi}{2}\right)$. Thus, the minimum energy configuration is this one. It is a uniformly continuous function of k and we have for all $i \ll_{i\to\infty} u_i(k) = m_i\pi$. At this limit, inequality (48-b) readily implies for all i, $|m_{i+1} + m_{i-1} - 2\, m_i|\pi \leq 2\pi$ or $b \leq 2$ which proves the first part of the theorem. (In general $b = 2$, since (50-a) implies that b is even. If $b = 0$, we necessarily have $m_i = 2pi + 2q$ with p and q integer. The corresponding configuration is a trivial ground-state $u_i = 2(pi+q)\pi$ with $l = 2p\pi$ and corresponds to a fixed point of the standard map).

When $\zeta = \dfrac{l}{2\pi}$ is irrational, $\{u_i\}$ is described by (45-a) and we have

$$m_i = 2F(il+\alpha) \tag{53-a}$$

with

$$F(x) = \mathrm{Int}\left(\frac{f(x)+\frac{\pi}{2}}{2\pi}\right) \tag{53-b}$$

$F(x)$ fulfills

$$F(x+2\pi) = F(x)+1 \tag{53-c}$$

Thus, if $F(x)$ has a discontinuity at x_0, it has the same discontinuities at $x_0 + 2n\pi$ where n is an arbitrary integer. Since $F(x)$ is monotonous increasing and only takes integer values, (53-c) implies that it has only one discontinuity with amplitude 1 in the interval $[0, 2\pi[$ and thus can be written

$$F(x) = \mathrm{Int}\left(\frac{x-x_0}{2\pi}\right) \tag{53-d}$$

Consequently, by rescaling the phase α, the coding sequence (53-a) takes the form (47-a).

Before continuing the proof of theorem 6, let us prove the following result which is a corollary of theorem 1:

PROPOSITION 1. *For $k > K(b)$ a sequence of stationary configurations $\{u_i^{(n)}\}$ with coding sequences $m_i^{(n)}$ fulfilling (8-a), is convergent iff (if and only if) the sequence of coding sequences $\{m_i^{(n)}\}$ is convergent.*

Note that this sequence cannot be uniformly convergent because of theorem 3.

Proof. Let us assume $\ll_{n\to\infty} \{u_i^{(n)}\} = \{u_i\}$. Since the set of minimum energy configurations Ω is closed for the weak topology [10], the accumulation points $\{u_i\}$ of minimum energy configurations $\{u_i^{(n)}\}$ are minimum energy configurations. Since $|u_i^{(n)} - m_i^{(n)}\pi| \leq \frac{\pi}{2}$, this property is preserved for $n \to \infty$ and thus $\{u_i\} \in E\left(\{m_i\};\frac{\pi}{2}\right)$ where $\{m_i\}$ is any accumulation point of the sequence

$\{m_i^{(n)}\}$. According to theorem 1, different coding sequences $\{m_i\}$ necessarily determine different stationary configurations. Therefore, $\{u_i\}$ belongs to a unique set $E\left(\{m_i\}; \frac{\pi}{2}\right)$ which proves that the sequence $\{m_i^{(n)}\}$ has a unique accumulation point $\{m_i\}$ and thus is convergent.

If we assume reciprocally that $\{m_i^{(n)}\}$ converge to $\{m_i\}$ (which fulfills (8-a)), since $|u_i^{(n)} - m_i^{(n)}\pi| \leq \frac{\pi}{2}$, the accumulation points of sequence $\{u_i^{(n)}\}$ belongs to $E\left(\{m_i\}; \frac{\pi}{2}\right)$ and are stationary configurations. Since theorem 1 predicts a unique stationary configuration in $E\left(\{m_i\}; \frac{\pi}{2}\right)$, $\{u_i^{(n)}\}$ converges to $\{u_i\}$.

□

We now go back to the proof of theorem 6. We consider coding sequences $\{m_i^{(n)}\}$ (47-a) with phase α_n and irrational $\frac{l_n}{2\pi}$ which converge to $\alpha \neq \frac{2p\pi}{s}$ (p integer) and to $\frac{l}{2\pi} = \frac{r}{s}$ rational respectively. Then, the coding sequences $\{m_i^{(n)}\}$ converges to $\{m_i\} = \left\{2\,\mathrm{Int}\left(\frac{ir}{s} + \frac{\alpha}{2\pi}\right)\right\}$ which fulfills (14-a). The limit minimum energy configuration $\{u_i\}$ corresponds to a s-periodic cycle of the standard map and thus is a commensurate ground-state. The set $\mathcal{G}_l$ is totally ordered as well as the associated set of coding sequences. When the phase α varies, the coding sequence $\{m_i\}$ with m_i even (see 52-a)) varies discontinuously at the phase $\alpha = \frac{2p\pi}{s}$ where p is any integer. Let us call $\{m_i^+\}$ and $\{m_i^-\}$, the two coding sequences at this discontinuity. We have

(54-a)	$m_i^+ - m_i^- = 2$	at the sites i fulfilling
(54-b)	$i = \iota$ modulo s ,	where ι fulfills
(54-c)	$\iota r + p = qs$	with q integer and $0 \leq \iota < s$ and
(54-d)	$m_i^+ - m_i^- = 0$	for $\iota \neq i$ modulo s

Since r and s are relatively prime, there is only one site ι per period s, where m_i changes. Consequently, there exists no coding sequence $\{m_i\}$ with even m_i and fulfilling (14-a) such that $\{m_i^+\} > \{m_i\} > \{m_i^-\}$. Thus, when the phase α varies, the coding sequences of all the commensurate ground-states of $\mathcal{G}_l$ are obtained by formula (47-a).

Finally, the coding sequences $\{m_i\}$ of discommensurations $\{v_i\}$ are easily found by using the inequality $\{u_p^+\} > \{v_i\} > \{u_i^-\}$ where $\{u_i^+\}$ and $\{u_i^-\}$ are "consecutive" commensurate ground-states in the totally ordered set $\mathcal{G}_l$ fulfilling (54). Thus we have

(55-a) $\quad m_i^+ = m_i^- = m_i \quad$ when $\iota \neq i$ modulo s

For an advanced discommensuration, condition (54-b) implies that

(55-b) $\quad m_{i+s} \geq m_i - 2r$

Then for $\iota = i$ modulo s or $i = ns + \iota$, there exists $i_0 = ns + \iota$ such that

(56-a) $$m_i^- = m_i < m_i^+ = m_i^- + 2 \qquad \text{for} \qquad i < i_0 \qquad \text{and}$$

(56-b) $$m_i^- < m_i = m_i^- + = m_i^- + 2 \qquad \text{for} \qquad i > i_0$$

A similar result hold for the delayed (or retarded) discommensurations.

□

In the anti-integrable regime, $k > K(2) > 2\sqrt{1+\pi^2}$, other results for the ground-state properties can be proven. According to a Lebesgue theorem [16], any monotonous increasing function $f(x)$ which can be associated with a Stieltjes measure $df(x)$, can be uniquely decomposed (up to a constant) into the sum of three monotonous increasing functions

(57) $$f(x) = f_{ac}(x) + f_{sc}(x) + f_d(x)$$

where $f_{ac}(b) - f_{ac}(a) = \int_a^b f'_{ac}(x)dx$ for any interval $[a, b]$. The component $f_{sc}(x)$ is a singular continuous function which means that $f_{sc}(x)$ is continuous function with a zero derivative $f'_{sc}(x)$ for most x. The continuous part of $f(x)$ is $f_c(x) = f_{ac}(x) + f_{sc}(x)$.

The discrete component $f_d(x)$ can be written as a convergent series of jump functions $Y(x)$ i.e. the positive measure $df_d(x)$ is a countable series of Dirac measures $\sum_i f_i\delta(x - x_i)$ with positive amplitudes f_i and locations x_i. This function only varies by discontinuities. Theorem 5 in ref. 13 asserts that close to the anti-integrable limit, the hull function $f(x)$ only contains a discrete (or atomic) part and thus no continuous part:

THEOREM 7. *For $k > K(2) = 4\sqrt{1+\pi^2}$, the hull function $f(x)$ of any incommensurate ground-state defined by (55-a) is purely discrete (i.e. defines a purely atomic Stieltjes measure). Thus, $f(x)$ is discontinuous and all the discontinuities points of $f(x)$ can be generated from a single one x_0 by*

(58-a) $$x_{n,m} = x_0 + nl + 2m\pi$$

where n and m are arbitrary integers. For any n, the amplitude of the discontinuities

(58-b) $$\delta_n = f^+(x_{n,m}) - f^-(x_{n,m}) > 0$$

which does not depend on m, are strictly positive.

In fact this theorem can be extended to a much wider class of models with weaker assumptions [14]. We just require that the Gap Parameter of the ground-state be non-vanishing in order that the hull function of an incommensurate ground state be purely discrete. However, for the FK model close to the anti-integrable limit, this theorem proves in addition the existence of a unique class of discontinuity which means that all discontinuity points can be generated from a single one x_0 by (58-a).

Proof. The hull function $f(x)$ fulfills the functional equation

$$f(x+l)+f(x-l)-2f(x)=k\sin\ f(x) \tag{59-a}$$

when $k>2\sqrt{1+\pi^2}$ and a fortiori for $k>K(2)=4\sqrt{1+\pi^2}$, eq.(51-b) implies for all x

$$\cos\ f(x)>\cos\ x_0=\frac{1}{k}\ \sqrt{k^2-4\pi^2}>0 \tag{59-b}$$

where x_0 is given by (49-b). Otherwise, since $g(x)=f(x)-x$ is a 2π periodic function, the variation of the continuous part of $f(x)$ over a period 2π, $\int_a^{a+2\pi} df_c(x)$ does not depend on the origin of this period 2π. Thus,

$$\int_a^{a+2\pi}(df_c(x+l)-df_c(x))=0 \tag{60-a}$$

and a fortiori for the left member of (59-a)

$$\int_a^{a+2\pi}\Big((df_c(x+l)-df_c(x))-(df_c(x)-df_c(x-l))\Big)=0 \tag{60-b}$$

which yields by integration of the continuous part of the right member

$$k\int_a^{a+2\pi}\cos\ f(x)df_c(x)=0 \tag{60-c}$$

Because of (59-b), the positive measure $df_c(x)$ has to be strictly zero. For $k>K(2)$, and for any coding sequence $\{m_i\}$ fulfilling (8-a) with $b\le 2$, the stationary configurations in $E\left(\{m_i\};\frac{\pi}{2}\right)$ is unique. Consequently, when varying the phase, the ground-states which are determined by the coding sequences (47-a), have discontinuities which just correspond to those of the sequence $\{m_i(\alpha)\}$. Then, it is clear that the discontinuity points of $f(x)$ are determined from a single one by (58-a). Since $g(x)=f(x)-x$ is 2π-periodic and $f(x)$ monotonous increasing, (58-b) is fulfilled.

□

The sequence $\{\delta_n\}$ has been interpreted as the bond modulation $\{u_{i+1}-u_i\}$ of the Frenkel-Kontorowa model associated with *an effective discommensuration*:

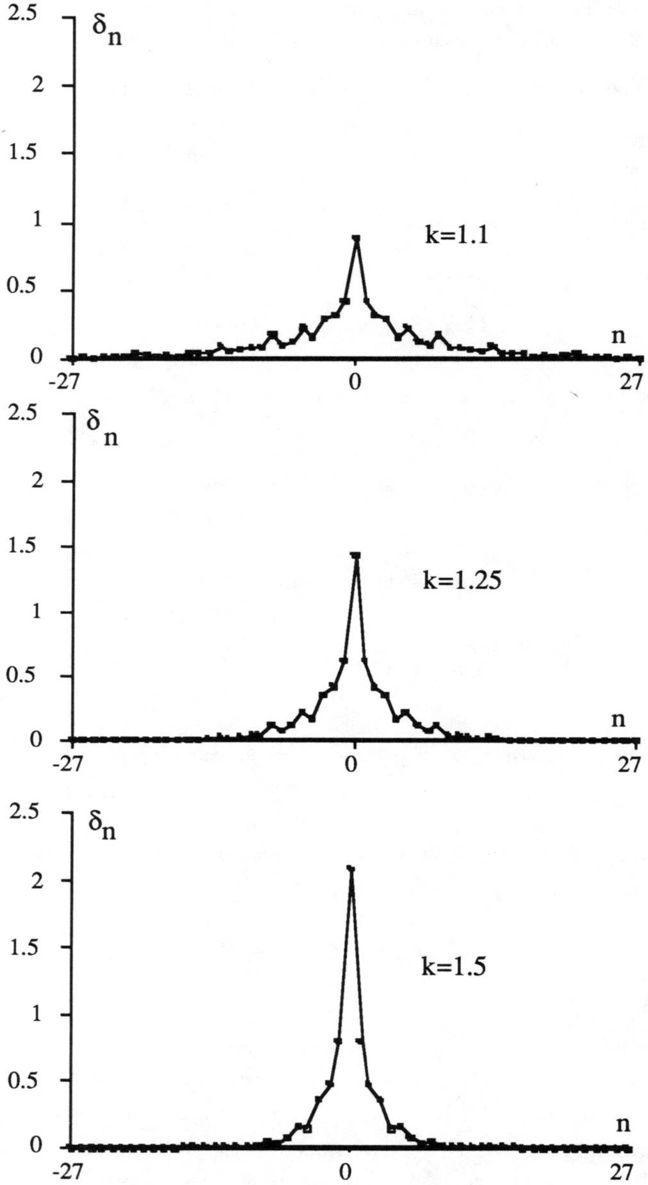

FIGURE 4: *Numerical calculation of* $\{\delta_n\}$ *defined by (62-a) versus n for several values of k in the FK model for* $\frac{l}{2\pi} = \frac{\sqrt{5}-1}{2}$. *Note that* $\sum_n \delta_n = 2\pi$. *We observe* $\{\delta_n\} \to 0$ *when* $k \to k_c\left(\frac{l}{2\pi}\right) \cong 0.97$

(61-a) $$u_{i+1} - u_i = f((i+1)l + \alpha) - f(il + \alpha)$$

Using theorem 7, we have

(61-b)

$$f(x+l) - f(x) = \sum_{n=-\infty}^{n=+\infty} \delta_n \left(\mathrm{Int}\left(\frac{x - x_0 - (n-1)l}{2\pi} \right) - \mathrm{Int}\left(\frac{x - x_0 - nl}{2\pi} \right) \right)$$
$$= \sum_{n=-\infty}^{n=+\infty} \delta_n \chi(x - x_0 - nl) \; .$$

For $0 \leq l < 2\pi, \chi(x) = \mathrm{Int}\left(\frac{x+l}{2\pi}\right) - \mathrm{Int}\left(\frac{x}{2\pi}\right)$ is 2π-periodic and equal to zero or one. Thus, we can write

(62-a) $$u_{i+1} - u_i = \sum_{n=-\infty}^{n=+\infty} \delta_n \chi(il + \alpha - x_0 - nl) = \sum_{n=-\infty}^{+\infty} \delta_{i+n} \sigma_n$$

where $\{\sigma_n\}$ is a quasi-periodic pseudo-spin sequence of 0 and 1

(62-b) $$\sigma_n = \chi(\alpha - x_0 - nl) \; .$$

We say that there exists a discommensuration at bond n when $\sigma_n = 1$ and none when $\sigma_n = 0$. The bond modulation (62-a) thus appears to be a *linear superposition* of effective discommensurations located at the occupied sites n (with $\sigma_n = 1$). The bond modulation associated with an effective discommensuration is of physical interest (Figure 4). This description also extends when $l \notin [0, 2\pi]$ except that the pseudo-spin sequence σ_n is built with two consecutive integers which are not 0 and 1.

We end this paper with another application of the concept of anti-integrability concerning the properties of the Average Action of the minimal orbits of the standard map versus their rotation number.

The average of the variational form (1) can be defined for each ground-state with commensurability ratio by

(63) $$\Psi(l) = \lim_{N-N' \to \infty} \frac{1}{N - N'} \sum_{N' \leq i < N'} L(u_{n+1}, u_n)$$

$\Psi(2\pi\omega) = A(\omega)$ in ref.17 is called Average Action. Results by J. Mather and V. Bangert [17,18] prove that in addition to be convex, $\Psi(l)$ is continuous for all irrational $\frac{l}{2\pi}$ and is discontinuous for rational $\frac{l}{2\pi} = \frac{r}{s}$ iff there exists no invariant circle which goes around the infinite cylinder $\mathbb{C} \times \mathbb{R} = \{u_i \mod 2\pi, p_i\}$ and consists entirely of periodic orbits with rotation number $\frac{r}{s}$. Since $\Psi(l)$ is convex, the right and left derivative of $\Psi(l)$ are defined for any l and are denoted $\Psi'^{+}(l)$ and $\Psi'^{-}(l)$

and $\Psi'^{-}(l)$ respectively. Among other conjectures, we claimed in ref.15 section 5 pp. 230 235, a conjecture (completeness of the Devil's staircase) which is equivalent to the following assertion:

If for $\zeta = \dfrac{l}{2a}$ *in some closed interval* I*, the Lyapounov coefficient of the associated minimal orbits is strictly positive and larger than a non zero positive number* γ*, then* $\Psi'^{+}(l)$ *(or equivalently* $\Psi'^{-1}(l)$*) is a purely discrete function of* $\zeta = \dfrac{l}{2\pi}$ *in* I*.*

Although we have not the full proof of this conjecture (except in some specific models associated with piecewise linear map [1]), we can prove the following theorem for the standard map close to its anti integrable limit:

THEOREM 8. *For the standard map with* $k > 2\sqrt{1+\pi^2}$*, the right derivative* $\Psi'^{+}(l)$ *and the left derivative* $\Psi'^{-}(l)$ *of the average action* $\Psi(l)$ *with respect to the atomic mean distance* l *(or rotation number* $\zeta = \dfrac{l}{2\pi}$*), are purely discrete functions (ie. their continuous part is a constant) and have discontinuities only for* $\dfrac{l}{2\pi} = \dfrac{r}{s}$ *rational.*

Proof. The proof closely follow the empirical arguments presented ref [15]). Close enough to the anti-integrable limit, the physicist's estimation can be transformed into exact bounds by taking advantage of the representation of the ground-state and minimum energy configurations by coding sequences (theorem 6).

The initial arguments consisted in approximating an incommensurate ground-state by piecewise discommensurations refereed to a given commensurate ground-state. Then, the difference between the average action of the incommensurate ground-state and those of the commensurate ground-state can be bounded.

However, the technical implementation of this proof is rather long. We organize this proof into several steps

1- We first establish a lemma which allows one to bound the distance between two minimum energy configurations in a finite domain.

2- We use this lemma for showing than an incommensurate ground-state can be approximated as an array of discommensurations refereed to an arbitrary commensurate ground-state within an accuracy which can be explicitly bounded.

3- We define the energy of discommensuration.

4- Using the inequalities obtained for the energy difference between an incommensurate ground-state and a commensurate ground-state as a function of the energy of these discommensurations and of their density within an accuracy which is bounded.

5- Finally, using the obtained inequality on this energy difference, we prove theorem 8, on the basis of a proposition already proven in ref. 15 (Appendix A2).

For performing this proof, we assume for sake of simplicity and without any lack of generality $0 \leq l < 2\pi$

Step 1: We prove the preliminary lemma

LEMMA. *For $k > 2\sqrt{1+\pi^2}$, let us consider two minimum energy configurations $\{u_i)$ and $\{v_i\}$ with coding sequences $\{m_i\}$ and $\{n_i\}$ respectively such that*

(64-a) $$m_i = n_i \qquad \textit{for} \qquad i_0 \le i \le i_1$$

Then we have

(64-b) $$|u_i - v_i| < \frac{A}{2}\,(\exp -\gamma|i-i_0| + \exp -\gamma|i-i_1|) \quad \textit{for} \quad i_0 \le i \le i_1$$

with

(65-a) $$\gamma = Ln\left(\frac{2+\sqrt{k^2-4\pi^2}}{2}\right) \qquad \text{and}$$

(65-b) $$A = \frac{\pi}{1-\exp(-\gamma)}$$

Proof of lemma. Conditions (49) and (50) used for the proof of theorem 6 yield that when $k > 2\sqrt{1+\pi^2}$, we have for any minimum energy configuration $\{u_i\}$

(66-a) $$|u_i - m_i\pi| \le x_0 = \text{Arcsin}\left(\frac{2\pi}{k}\right) < \frac{\pi}{2}$$

The two minimum energy configurations $\{u_i)$ and $\{v_i\}$ both fulfills eq.2-b which yields by difference

(66-b) $$(2+k\cos\ \xi_i)(u_i - v_i) - (u_{i+1} - v_{i+1}) - (u_{i-1} - v_{i-1}) = 0$$

where ξ_i is some number in the interval determined by u_i and v_i. Since u_i and v_i both fulfills the inequality (64-a) for $i_0 \le i \le i_1$, we have

(66-c) $$|\xi_i - m_i\pi| \le x_0 = \text{Arcsin}\left(\frac{2\pi}{k}\right) \qquad \text{for } i_0 \le i \le i_1$$

which implies

(66-d) $$2+\sqrt{k^2-4\pi^2} = 2 + k\ \cos\ x_0 \le 2 + k\ \cos\xi_i$$

For $i_0 < i < i_1$, equation (66-b) can be written in matrix form

(67-a) $$\overline{\overline{P}}\,\overline{U} = \overline{V}\ .$$

$\overline{\overline{P}}$ is a $(i_1 - i_0 - 1) \times (i_1 - i_0 - 1)$ matrix which has $\overline{\overline{D}}$ as diagonal part

(67-b) $$D_{i,j} = (2 + k\ \cos\ \xi_i)\delta_{i,j} \qquad \text{for } i_0 < i < i_1 \text{ and } i_0 < j < i_1$$

$\overline{\overline{F}}$ is the off diagonal part of $\overline{\overline{P}}$ with Hermitian norm $||\overline{\overline{F}}|| = 2$. $\overline{U}$ and $\overline{V}$ are $(i_1 - i_0 - 1)$ column vectors defined as

(67-c) $U_i = (u_i - v_i)$ for $i_0 < i < i_1$

(67-d) $V_i = 0$ for $i_0 + 1 < i < i_1 - 1$

(67-e) $V_{i_0} + 1 = (u_{i_0} - v_{i_0})$ and

(67-f) $V_{i_1 - 1} = (u_{i_1} - v_{i_1})$

For finding the bound (65-a), we analyze the decay rate of the coefficients $[\overline{\overline{P}}^{-1}]_{i,j}$ of the inverse matrix $\overline{\overline{P}}^{-1}$ as a function of $|i-j|$. Because of (66-c) and (67-b), we have

(68-a) $$||\overline{\overline{D}}^{-1}|| \leq \frac{1}{2 + \sqrt{k^2 - 4\pi^2}}$$

and

(68-b) $$||\overline{\overline{F}}.\overline{\overline{D}}^{-1}|| \leq \frac{2}{2 + \sqrt{k^2 - 4\pi^2}} = \exp(-\gamma) < 1$$

Thus we can write as a convergent series

(69-a) $$\overline{\overline{P}}^{-1} = \overline{\overline{D}}^{-1} \sum_{n=0}^{\infty} (-1)^n (\overline{\overline{F}}.\overline{\overline{D}}^{-1})^n$$

Since $F_{i,j} = 0$, for $|i-j| > 1$ and since $\overline{\overline{D}}^{-1}$ is diagonal, it comes out that

(69-b) $$[\overline{\overline{D}}^{-1}(\overline{\overline{F}}.\overline{\overline{D}}^{-1})^n]_{i,j} = 0 \qquad \text{for } |i-j| > n.$$

Consequently, we have

(69-c) $$[\overline{\overline{P}}^{-1}]_{i,j} = \sum_{n \geq |i-j|} (-1)^n [\overline{\overline{D}}^{-1}(\overline{\overline{F}}.\overline{\overline{D}}^{-1})^n]_{i,j}$$

This property implies

(70-a) $$|[\overline{\overline{P}}^{-1}]_{i,j}| \leq \sum_{n \geq |i-j|} ||\overline{\overline{D}}^{-1}(\overline{\overline{F}}.\overline{\overline{D}}^{-1})^n||$$

$$\leq \sum_{n \geq |i-j|} \frac{1}{2 + \sqrt{k^2 - 4\pi^2}} \exp(-n\gamma) = \frac{\exp(-\gamma)}{2} \frac{\exp(-\gamma|i-j|)}{1 - \exp(-\gamma)} \quad .$$

The solution of (67-a) yields for $i_0 < i < i_1$

(70-b)

$$(u_i - v_i) = \sum_{j=i_0+1}^{j=i_1-1} [\overline{\overline{P}}^{-1}]_{i,j} V_j$$

$$= [P^{-1}]_{i,i_0+1}(u_{i_0} - v_{i_0}) + [\overline{\overline{P}}^{-1}]_{i,i_1-1}(u_{i_1} - v_{i_1})$$

The proof of (64-b) is now straightforward by using (70-a) and (70-b), and noting that $|u_{i_0}| < \pi$ and $|u_{i_1} - v_{i_1}| < \pi$.

□

Step 2: Let us now decompose an incommensurate ground-state into an array of discommensurations referred to an arbitrary ground-state.

The coding sequences $\{m_i\}$ of the commensurate ground-states $\{u_i\}$ with commensurability ratio $\frac{l}{2\pi} = \frac{r}{s}$ (with r and s irreducible) have the form (theorem 6).

(71-a) $$m_i = 2\,\mathrm{Int}\left(\frac{ir}{s} + \alpha\right)$$

where α is an arbitrary phase. The coding sequence and thus the commensurate ground-state is the same for any

(71-b) $$\frac{p}{s} \leq \alpha < \frac{p+1}{s} \qquad \text{where } p \text{ is an integer}$$

We denote $\{u_i^{(p)}\}$ the commensurate ground-state with a phase α fulfilling (71-b). Transformation $T_{n,m} : \{u_i\} \to \{u_{i+n} - 2\pi m\}$ with n and m integer, changes the phase α of $\{u_i\}$ into $\alpha + n2\pi\frac{r}{s} - 2\pi\, m = \alpha + 2\pi\frac{nr - ms}{s}$ (see ref.10). Applying these transformations $T_{n,m}$ for all n and m to a single commensurate ground-state generates the whole set of commensurate ground-states $\mathcal{G}_{2\pi r/s}$ which is totally ordered. It will be useful in the next, to note that with the same proof as for the incommensurate ground-states [10], the whole set of commensurate ground-states can be described with a single hull function $f_c(x)$ which is monotonous increasing and with $g_c(x) = f_c(x) - x$, 2π-periodic. We have

(71-c) $$u_i^{(p)} = f_c\left(2\pi\frac{ir + p}{s}\right)$$

Because of (71-b), $f_c(x)$ is piecewise constant on the intervals $\frac{p}{s} \leq x < \frac{p+1}{s}$ with p integer.

Let us now consider an incommensurate ground state $\{v_i\}$ with commensurability ratio $\zeta = \frac{l}{2\pi}$. In order to fix the ideas we assume $\frac{l}{2\pi} > \frac{r}{s}$. This ground-state has the coding sequence

(72-a) $$n_i = 2\,\mathrm{Int}\left(\frac{il}{2\pi}\right) = 2\,\mathrm{Int}\left(\frac{ir}{s} + \alpha_i\right)$$

with

(72-b) $$\alpha_i = i\left(\frac{l}{2\pi} - \frac{r}{s}\right)$$

It is convenient to define a sequence of integers $\{i_p\}$

(73-a) $$i_p = \text{Int}\left(\frac{2\pi p}{sl - 2\pi r}\right) + 1$$

in order to have

(73-b) $$\frac{p}{s} \le \alpha_i < \frac{p+1}{s} \qquad \text{for} \qquad i_p \le i < i_{p+1}$$

and the sequence of integers $\{j_p\}$ which determine the cutting of the incommensurate ground-state into advanced discommensurations (cf.fig.5)

(74-a) $$j_p = \text{Int}\left(\frac{(2p-1)\pi}{sl - 2\pi r}\right) + 1$$

in order to have

(74-b) $$\frac{p - \frac{1}{2}}{s} \le \alpha_i < \frac{p + \frac{1}{2}}{s} \qquad \text{for} \qquad j_p \le i < j_{p+1}$$

For $i_{p-1} \le i < i_{p+1}$, the coding sequence $\{n_i\}$ is identical to the coding sequence of the advanced discommensuration $\{w_i^{(p)}\}$ defined by (47-b) with $n = i_p$. Thus we can use the above lemma for bounding the distance

(75) $$|v_i - w_i^{(p)}| \ < \frac{A}{2} \left(\exp -\gamma|i - i_{p-1}| + \exp -\gamma|i + 1 - i_{p+1}|\right)$$

For the restricted interval, $j_p \le i \le j_{p+1}$, we have using the definitions (73-a) and (74-a)

(76-a) $$|i - i_{p-1}| \ \ge |j_p - i_{p-1}| \ \ge \text{Int}\left(\frac{\pi}{sl - 2\pi r}\right) \ge \frac{\pi}{sl - 2\pi r} - 1$$

and

(76-b) $$|i - 1 - i_{p+1}| \ \ge |j_{p+1} + 1 - i_{p+1}| \ \ge \text{Int}\left(\frac{\pi}{sl - 2\pi r}\right) - 1 \ge \frac{\pi}{sl - 2\pi r} - 2$$

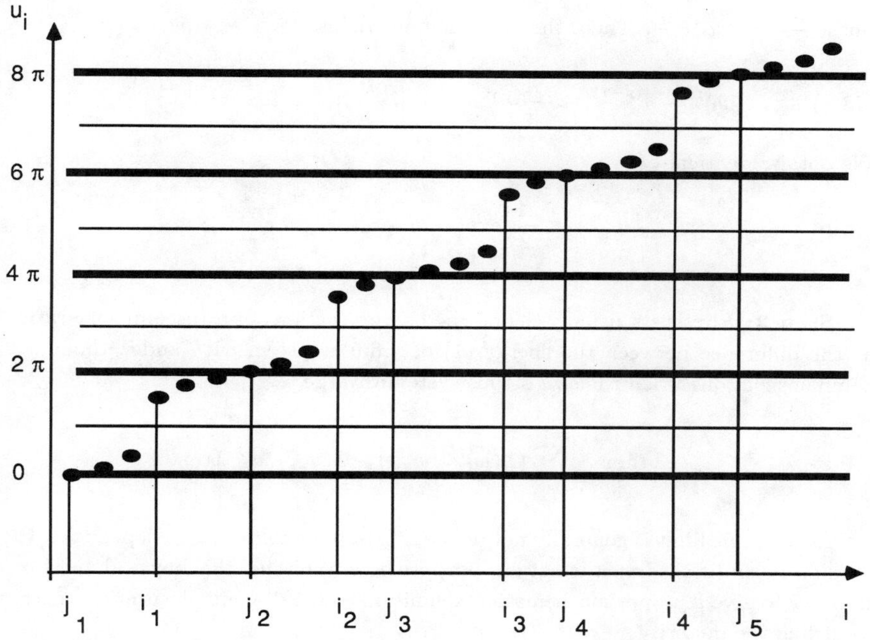

FIGURE 5: *Scheme showing the cutting of an incommesurate structure into (approximate) discommensurations refereed to the commensurate structure $l = 0$. Discommensurations are delimited by two consecutive black lines that is by $j_n \leq i < j_{n+1}$*

Therefore, it comes out from (75)

$$|v_i - w_i^{(p)}| < A\ e^{2\gamma} \exp\left(-\frac{\gamma}{2s}\ \frac{1}{\dfrac{l}{2\pi} - \dfrac{r}{s}}\right) \quad \text{for} \quad j_p \leq i \leq j_{p+1} \tag{77}$$

As it was conjectured in ref.15, on the basis of physical arguments, it appears that the approximation of an incommensurate ground-state by piecewise discommensuration becomes very accurate when the rotation number $\frac{l}{2\pi}$ of the incommensurate ground-state goes to the rational rotation number $\frac{r}{s}$ of the commensurate ground-state.

Considering now the discommensuration $\{w_i^{(p)}\}$ and using (47-b), for all $i < i_p$, it is the same coding sequence as the commensurate ground-state $\{u_i^{(p-1)}\}$ with phase $\frac{p-1}{s} \le \alpha < \frac{p}{s}$. Then, the above lemma yields:

(78-a) $$|w_i^{(p)} - u_i^{(p-1)}| < \frac{A}{2} e^{\gamma} \exp(-\gamma|i - i_p|) \qquad \text{for} \quad i < i_p .$$

We obtain identically

(78-b) $$|w_i^{(p)} - u_i^{(p)}| < \frac{A}{2} \exp(-\gamma|i - i_p|) \qquad \text{for} \quad i \geq i_p$$

Step 3: Physicists usually define the energy (action) of a discommensuration as the difference between the energy (1) of a finite system with and without discommensurations when the size of this system diverge:

(79-a) $$\lim_{N'-N\to\infty} \sum_{i=N}^{i=N'} (L(w_{i+1}^{(p)}, w_i^{(p)}) - L(u_{i+1}^{(p)}, u_i^{(p)}))$$

In fact, this limit is generally not defined!!!. In physical terms, it depends on the boundary conditions but it is not our purpose here to discuss this physical question. In order to give a proper mathematical definition to the discommensuration energy, we define the quantity

(79-b)
$$e_m^+ = \sum_{i=-\infty}^{i=m-1} \left(L(w_{i+1}^{(1)}, w_i^{(1)}) - L(u_{i+1}^{(0)}, u_i^{(0)}) \right) + \sum_{i=m}^{i=+\infty} \left(L(w_{i+1}^{(1)}, w_i^{(1)}) - L(u_{i+1}^{(1)}, u_i^{(1)}) \right)$$

as the sum of two series which are absolutely convergent because of (78). This sum depends on m.

For the commensurate ground-state, the sequence $L_i^{(p)} = L(u_{i+1}^{(p)}, u_i^{(p)})$ has period s : $L_{i+s}^{(p)} = L_i^{(p)}$. Thus, the energy per atom (average action)

(80-a) $$\Psi\left(2\pi\frac{r}{s}\right) = \frac{1}{s} \sum_{i=m}^{i=m+s-1} L(u_{i+1}^{(p)}, u_i^{(p)})$$

is obtained from the average over any unit cell of s consecutive atoms for any commensurate ground-state. Particularly, we have

(80-b) $$\sum_{i=m}^{i=m+s-1} L(u_{i+1}^{(0)}, u_i^{(0)}) = \sum_{i=m}^{i=m+s-1} L(u_{i+1}^{(1)}, u_i^{(1)})$$

which implies

(80-c) $$e_{m+s}^+ = e_m^+$$

By inspection of (47-b), it comes out that the coding sequence $\{m_i\}$ for a discommensuration located at site n remains the same for n such that

(81-a) $$n_0 + js < n \leq n_0 + (j+1)\, s \qquad , j \text{ integer}$$

where $0 \leq n_0 < s$ is the solution of

(81-b) $$(n_0 r + p) \text{ modulo } s = s - 1$$

Thus the advanced discommensuration $\{w_i^{(1)}\}$ is unchanged for i_1 in some interval of width s. Thus, because of (80-c) that e_m^+ does not depend on the location of the discommensuration i_1. The other advanced discommensurations can be obtained from the discommensuration $w_i^{(0)}$ by transformations $T_{n,m}$. More precisely, we have

(82-a) $$w_i^{(p)} = w_{i+n_p}^{(0)} - 2\pi 1_p$$

where the integers n_p and l_p fulfills the integer equation

(82-b) $$n_p r - 1_p s = p$$

Using this transformation, it comes out

(82-c) $$\sum_{i=-\infty}^{i=m-1} \left(L(w_{i+1}^{(p)}, w_i^{(p)}) - L(u_{i+1}^{(p-1)}, u_i^{(p-1)}) \right)$$
$$+ \sum_{i=m}^{i=+\infty} \left(L(w_{i+1}^{(p)}, w_i^{(p)}) - L(u_{i+1}^{(p)}, u_i^{(p)}) \right) = e_{m+n_{p-1}}^+$$

The average

(83) $$e^+\left(2\pi\frac{r}{s}\right) = \frac{1}{s}\sum_{m=1}^{s} e_m^+$$

is physically the energy of an advanced discommensuration refereed to the commensurate ground-state with commensurability ratio $\frac{r}{s}$.

Step 4: We use the above result for proving the following proposition which bounds the difference $\Psi(l) - \Psi\left(2\pi\frac{r}{s}\right)$.

PROPOSITION 2. *Let us consider* $0 \leq 2\pi\frac{r}{s} < l \leq 2\pi$, *the average action fulfills the inequality*

(84-a)
$$|\Psi(l) - \Psi\left(2\pi\frac{r}{s}\right) - s|\frac{l}{2\pi} - \frac{r}{s}|e^+\left(2\pi\frac{r}{s}\right)|$$
$$< A\,(2\pi + k)\left(2\, e^{2\gamma} + s|\frac{l}{2\pi} - \frac{r}{s}|\frac{1}{1-e^{-\gamma}}\right) \exp\left(-\frac{\gamma}{2s}\frac{1}{|\frac{l}{2\pi} - \frac{r}{s}|}\right) \quad .$$

The same inequality hold for $0 \leq l < 2\pi\frac{r}{s} \leq 2\pi$ *but with the energy* $e^-\left(2\pi\frac{r}{s}\right)$ *of a delayed discommensuration. A consequence obtained by taking* $l \to 2\pi\frac{r}{s}$, *is*

(84-b) $$\Psi'^+\left(2\pi\frac{r}{s}\right) = \frac{s}{2\pi}\, e^+\left(2\pi\frac{r}{s}\right)$$

(84-c) $$\Psi'^-\left(2\pi\frac{r}{s}\right) = -\frac{s}{2\pi}\, e^-\left(2\pi\frac{r}{s}\right)$$

Proof. We first compare the energy of the incommensurate ground-state for a finite system. We first have

(85-a)
$$L(v_{i+1}, v_i) - L(w_{i+1}^{(p)}, w_i^{(p)}) = (w_{i+1}^{(p)} - v_{i+1})\partial_1 L(\omega_{i+1}, \omega_i) + (w_i^{(p)} - v_i)\partial_2 L(\omega_{i+1}, \omega_i)$$

where $\omega_{i+1} = v_{i+1} + \theta(w_{i+1}^{(p)} - v_{i+1})$ and $\omega_i = v_i + \theta\,(w_i^{(p)} - v_i)$ and θ is some number between 0 and 1. For $0 \leq l < 2\pi$, the derivatives $\partial_1 L(\omega_{i+1}, \omega_i)$ and $\partial_2 L(\omega_{i+1}, \omega_i)$ are easily bounded by $(2\pi + k)$ which yields for $j_p \leq i \leq j_{p+1}$ using (77)

(85-b) $$|L(v_{i+1}, v_i) - L(w_{i+1}^{(p)}, w_i^{(p)})| < 2(2\pi + k)Ae^{\gamma}\exp\left(-\frac{\gamma}{2s}\frac{1}{\frac{l}{2\pi} - \frac{r}{s}}\right)$$

Thus, the average energy for the finite incommensurate ground-state with $j_N - j_1$ atoms is

(86-a)
$$F_N = \frac{1}{j_N - j_1}\sum_{j_i < i < j_N} L(v_{i+1}, v_i) = \frac{1}{j_N - j_1}\sum_{p=1}^{p=N-1}\left(\sum_{j_p \leq i < j_{p+1}} L(v_{i+1}, v_i)\right)$$
$$= F'_N + R_N$$

where

(86-b) $$F'_N = \frac{1}{j_N - j_1}\sum_{p=1}^{p=N-1}\left(\sum_{j_p \leq i < j_{p+1}} L(w_{i+1}^{(p)}, w_i^{(p)})\right)$$

is the average energy for the (non-stationary) configuration of piecewise discommensurations and

(86-c) $$|R_N| < 2\,(2\pi + k)A\,e^{\gamma}\exp\left(-\frac{\gamma}{2s}\frac{1}{\frac{l}{2\pi} - \frac{r}{s}}\right)$$

is the remainder which is bounded.

We can compare each term in (86-b) with a single discommensuration energy contribution (79-b) or (82-c). The energy contribution which comes from the outside of the interval $[j_p, j_{p+1}]$ can be bounded by proceeding as in (85) and using (76-a) and (78)

(87-a)

$$|\sum_{i=-\infty}^{i=j_p-1}\left(L(w_{i+1}^{(p)}, w_i^{(p)}) - L(u_{i+1}^{(p-1)}, u_i^{(p-1)})\right)|$$

$$\leq (2\pi + k)\frac{A}{2}e\gamma \sum_{i=-\infty}^{j_p-1} \exp(-\gamma(i_p - i)) = (2\pi + k)\frac{A}{2}\frac{\exp -\gamma|i_p - j_p|}{1 - e^{-\gamma}}$$

(87-b) $$|\sum_{i=j_{p+1}}^{i=+\infty}\left(L(w_{i+1}^{(p)}, w_i^{(p)}) - L(u_{i+1}^{(p)}, u_i^{(p)})\right)| \leq (2\pi + k)\frac{A}{2}\frac{\exp -\gamma|i_p - j_{p+1}|}{1 - e^{-\gamma}}$$

Thus, we can write

(88-a)

$$\sum_{j_p \leq i < j_{p+1}}\left(L(w_{i+1}^{(p)}, w_i^{(p)}) - L(u_{i+1}^{(p-1)}, u_i^{(p-1)})\right)$$

$$+ \sum_{i=i_p}^{i=j_{p+1}-1}\left(L(w_{i+1}^{(p)}, w_i^{(p)}) - L(u_{i+1}^{(p)}, u_i^{(p)})\right) = e^{+}_{i_p+n_{p-1}} + \eta_p$$

where using (87), we have for the remainder

(88-b) $$|\eta_p| < (2\pi + k)\frac{A}{2}\frac{\exp -\gamma|i_p - j_p| + \exp -\gamma|i_p - j_{p+1}|}{1 - e^{-\gamma}}$$

$$< (2\pi + k)\frac{A}{1 - e^{-\gamma}} \exp -\gamma\left(\frac{\pi}{sl - 2\pi r}\right) .$$

Then, the average energy of the piecewise discommensuration configuration fulfills

(89-a) $$(j_n - j_1)F'_N = \sum_{p=1}^{N-1}(e^{+}_{i_p+n_{p-1}} + \eta_p) + (j_n - j_1)F''_N$$

where

(89-b) $$F''_N = \frac{1}{j_N - j_1}\sum_{p=1}^{N}\left(\sum_{j_p \leq i < i_p} L(u_{i+1}^{(p-1)}, u_i^{(p-1)}) + \sum_{i_p \leq i < j_{p+1}} L(u_{i+1}^{(p)}, u_i^{(p)})\right)$$

is the average energy of a (non-stationary) configuration which consists in pieces of commensurate ground-states with different phases. For calculating this average energy, we define the 2π-periodic function

(90-a) $$L(x) = L(f_c\left(x + 2\pi\frac{r}{s}\right), f_c(x))$$

with the hull function (71-c). It is also piecewise constant in the same intervals $\frac{p}{s} \le x < \frac{p+1}{s}$. The average energy of the commensurate ground-state is

(90-b) $$\Psi\left(2\pi\frac{r}{s}\right) = \frac{1}{s}\sum_{i=1}^{s} L(u_{i+1}^{(0)}; u_i^{(0)}) = \frac{1}{2\pi}\int_0^{2\pi} L(x)dx$$

Because of definition (73b), the sequence which appears in (89-b) defined as

(91-a)
$$L_i = L(u_{i+1}^{(p)}, u_i^{(p)}) = L\left(f_c\left(2\pi\frac{(i+1)r+p}{s}\right), f_c\left(2\pi\frac{ir+p}{s}\right)\right)$$
$$\text{for} \qquad i_p \le i < i_{p+1}$$

can be written with this hull function $L(x)$ as

(91-b) $$L_i = L\left(f_c(2\pi(i+1)\frac{r}{s} + 2\pi\alpha_i\right), \left(f_c\left(2\pi i\frac{r}{s} + 2\pi\alpha_i\right)\right) = L(il)$$

where α_i is given by (72-b). For $\frac{l}{2\pi}$ irrational, we have because of (89-b)

(91-c) $$\lim_{N\to\infty} F_N'' = \lim_{N\to\infty}\frac{1}{j_N - j_1}\sum_{j_1\le i<j_N} L(il) = \Psi\left(2\pi\frac{r}{s}\right)$$

Turning back to the expression (89-a), we have to estimate the average of $e^+_{i_p+n_{p-1}}$. Since we have $n_{p+s} = n_p$, we first consider the average of the sequence

(92-a) $$a_n = e^+_{i_{p+ns}+n_{p-1+ns}} = e^+_{i_{p+ns}+n_{p-1}}$$

The sequence $l_n = i_{p+ns}$ modulo s = Int $\left(\frac{2\pi(p+ns)}{sl-2\pi r}\right) + 1$ modulo s = $\chi\left(\frac{2\pi\ n}{sl-2\pi r}\right)$ defined by (73-a) can be generated by a rotation on a unit circle with incommensurate angle $\frac{2\pi}{sl-2\pi r}$ where $\chi(x)$ is 1 periodic and piecewise constant on equal intervals of width $\frac{1}{s}$. Since the sequence $\frac{2\pi\ n}{sl-2\pi r}$ modulo 1 is uniformly distributed on the circle, the sequence i_{p+ns} modulo s over n, is uniformly distributed on the numbers from 1 to s. Therefore, the average of a_n is just the discommensuration energy (83) and does not depend on p. By taking the limit $N \to \infty$ in (86-a) and (89-a) and since the density of advanced discommensurations is

(92-b) $$\rho_{r/s}(l) = \lim_{N\to\infty}\frac{N-1}{j_N - j_1} = s\left(\frac{l}{2\pi} - \frac{r}{s}\right)$$

inequality (84-a) is proven for $0 \le \frac{2\pi r}{s} < l \le 2\pi$. The proof is valid by continuity for any l in this interval. A similar proof holsd for $l < \frac{2\pi r}{s}$. □

Step 5: The end of the proof consists in using inequality (84-a) at all rational $\frac{l}{2\pi}$. Since $\Psi(l)$ is a convex function, proposition 2 implies

$$0 < \Psi(l) - \Psi(l_c) - (l - l_c)\Psi'^{+}(l_c) < h_s(l - l_c) \tag{93-a}$$

with

$$h_s(x) = A\ (2\pi + k)\left(2\ e^{2\gamma} + 2\pi s\ x\frac{1}{1 - e^{-\gamma}}\right)\exp\left(-\frac{\gamma}{2s}\,\frac{2\pi}{x}\right) \tag{93-b}$$

for any rational $\frac{l_c}{2\pi} = \frac{r}{s}$ with r and s irreducible and $0 \leq l_c < l \leq 2\pi$ where a, b and γ are positive non vanishing constants which are independent of l , l_c and s. Of course a similar inequality hold for $0 \leq l < l_c \leq 2\pi$.

For proving theorem 8, we prove that the variation of the monotonous increasing function $\Psi'^{+}(l)$ over the set I_∞ of irrational $\frac{l}{2\pi}$ is zero. This result can be expressed as a more general proposition (Appendix A2 of ref.15) which just requires that the series $s\ h_s\left(\frac{1}{s^{1+\nu}}\right)$ for some positive ν be absolutely convergent. We have

Proposition 3. *Let $\Psi(x)$ be a convex function of x on the interval $[0, 2\pi]$. Let us assume that for all rational $\frac{x_c}{2\pi} = \frac{r}{s}$ (r and s irreducible) and $x > x_c$, we have*

$$0 < \Psi(x) - \Psi(x_c) - (x - x_c)\Psi'^{+}(x_c) < h_s(x - x_c) \tag{94-a}$$

where $h_s(x)$ are functions fulfilling the condition of convergence

$$\sum_{s=1}^{\infty} s\ h_s\left(\frac{2\pi}{s^{1+\nu}}\right) < \infty \tag{94-b}$$

for some number $0 < \nu < 1$.
Then, $\Psi'^{+}(x)$ is a purely discrete function, with all its discontinuities at $x_c = 2\pi\frac{r}{s}$.

Proof. First, we define the union of intervals

$$I_S^\nu(\xi) = \bigcup_{s>S}\left[2\pi\frac{r}{s}\ ,\ 2\pi\left(\frac{r}{s} + \frac{\xi}{s^{1+\nu}}\right)\right[\tag{95-a}$$

which depends on the integer S and on $0 < \xi < 1$. It is well-known that for any irrational number ζ, there exists a series of rational $\frac{p_i}{q_i}$ which converges to ζ and such that $0 < \zeta - \frac{p_i}{q_i} < \frac{1}{q_i^2}$ (Asymmetric Hurwitz theorem [20]). Thus, this property implies the weaker condition that there exists for any $\xi > 0$ and $0 < \nu < 1$, a subsequence of this series $\frac{p'_i}{q'_i}$ such that $0 < \zeta - \frac{p'_i}{q'_i} < \frac{\xi}{q_i'^{1+\nu}}$. As a result we have

$$I_S^\nu(\xi) \supset I_\infty \tag{95-b}$$

The variation $\mathrm{Var}(\Psi'^{+}(x) : I_\infty)$ of $\Psi'^{+}(x)$ over I_∞ fulfills

(96-a)
$$0 \leq \mathrm{Var}(\Psi'^{+}(x); I_\infty) \leq \mathrm{Var}(\Psi'^{+}(x); I_S^\nu(\xi))$$
$$\leq \sum_{s>S, 0<\frac{r}{s}<1} \left(\Psi'^{+}\left(2\pi\left(\frac{r}{s}+\frac{\xi}{s^{1+\nu}}\right)\right) - \Psi'^{+}\left(2\pi\frac{r}{s}\right)\right)$$

since $\Psi'^{+}(x)$ is monotonous increasing. Since (96-a) is true for any $0 < \xi < 1$. We can average over ξ. It comes after intergation over ξ.

(96-b)
$$0 \leq \mathrm{Var}(\Psi'^{+}(x); I_\infty)$$
$$\sum_{s>S, 0<\frac{r}{s}<1} \left[\Psi\left(2\pi\left(\frac{r}{s}+\frac{1}{s^{1+\nu}}\right)\right) - \Psi\left(2\pi\frac{r}{s}\right) - \frac{2\pi}{s^{1+\nu}}\Psi'^{+}\left(2\pi\frac{r}{s}\right)\right]\frac{s^{1+\nu}}{2\pi}$$

The form of (96-b) is appropriate for using the inequality (94-a). We have

$$0 \leq \mathrm{Var}(\Psi'^{+}(x); I_\infty) \leq \sum_{s>S, 0<\frac{r}{s}} h_s\left(\frac{2\pi}{s^{1+\nu}}\right) \leq \sum_{s>S} s\, h_s\left(\frac{2\pi}{s^{1+\nu}}\right) \tag{97-a}$$

since there is at most s irreducible rationals in the interval [0,1] with denominator s. Because of (94-b), the right member of (97-a) is the remainder of a convergent series at order S. It goes to zero $S \to \infty$. Consequently, we have

$$\mathrm{Var}(\Psi'^{+}(l); I_\infty) = 0 \tag{97-b}$$

which finally proves proposition 3 and theorem 8.

□ □

6. Conclusion. In summary, we have shown in this proceeding some applications of the concept of anti-integrability which were essentially focused on the standard map. By contrast with the KAM theory, the basic idea associated with the concept of anti-integrability, consists in expanding the model around the limit where it has a full chaotic behavior (anti-integrable limit). Then, it is shown that the chaotic trajectories are preserved under small perturbations providing conditions on their coding sequence, and that they can be followed continuously. When the perturbation parameter increases (smaller k), most of these chaotic trajectories have to disappear after bifurcations. We can also take advantage of the vicinity of the anti-integrable limit for finding extra results concerning the minimal orbits associated with the ground-states of the Frenkel–Kontorowa model and also concerning the property of the average action of these orbits as a function of the rotation number. A number of important questions are still open concerning for example the global measure of the chaotic trajectories in the standard map. It might be that this new approach could help for finding new results.

Although we essentially focused here on the standard map, the concept of anti-integrability is not restricted to this model. It can be easily extended not only to similar models [14] but it can be used in many other different dynamical models including non-area preserving maps and even circle maps [6]. It has also many potential applications in non-dynamical variational problems and for example in adiabatic models with coupled electrons and phonons [20] were this method yields rigorous proof for the existence of the bipolaronic states (which was still controversed in solid state physics).

Acknowledgement. A am indebted for many useful discussions while attending at the workshop with Chris Golé, Michael Herman, Stephane Laederich, Rafaël de la Llave, Robert MacKay, Jurg Moser, Tom Spencer and many other participants who kindly answers to my questions. I also thank John Mather for sending me preprints prior to publications and for comments.

This work has been supported in part by the EEC project: SCI*/0229-C (AM).

REFERENCES

[1] S. AUBRY, *The new concept of Transition by Breaking of Analyticity in a Crystallographic Model* in " *Solitons and Condensed Matter Physics*", Solid State Sciences **8** (1978), pp. 264–277 (Springer) ed A.R. Bishop and T. Schneider.

[2] I. PERCIVAL, J. Phys, **A12** (1979) L57.

[3] C. RADIN, *Low Temperature and the Origin of Crystalline Symmetry*, Int. J. Mod. Phys. **B1** (1987), pp. 1157–1191 and *Disordered Ground States of Classical Lattice Models* Preprint (1990).

[4] For the Banach fixed point theorem see for example E. Zeidler "*Nonlinear Functional Analysis and its Applications I.Fixed-point Theorems*" Theorem 1.A p. 17 and Proposition 1.2 p. 19. Springer (1986).

[5] B.V. CHIRIKOV, Phys. Rep.**52** 263 (1979).

[6] S. AUBRY AND G. ABRAMOVICI, *Chaotic Trajectories in the Standard Map, the Concept of Anti-Integrability*, Physica **D43** (1990), 199–219.

[7] S. AUBRY, *Structures Incommensurables et Brisure de la Symétrie de Translation in Structures et Instabilités*, Ed. C. Godreche Editions de Physique, France (1986), pp. 73–194.

[8] M. HERMAN, *Dynamics on Lagrangian Tori Invariant by Symplectic Diffeomorphisms*, see this proceeding.

[9] R. GRIFFITHS, private communication.

[10] S. AUBRY AND P.Y. DE DAERON, *The Discrete Frenkel–Kontorowa Model and its Extensions-Exact Results for the Ground-State*, Physica **D8** 381–422 (1983).

[11] J. MATHER, *Existence of Quasi-periodic Orbits for Twist Homeomorphism of the Annulus Topology*, **21** (1982), pp. 457–468.

[12] S. AUBRY, *The Twist Map, the Extended Frenkel–Kontorowa Model and the Devil's Staircase*, Physica **D7** (1983), pp. 240–258.

[13] S. AUBRY, *Analyticity Breaking and Anderson Localization in Incommensurate Lattices* in *Proceeding of the VIII International Colloquium on Group Theoretical Methods in Physics*, Ed, L.P. Horwitz and Y. Ne'man, Ann. Israel Phys. Soc. **3** (1980) pp. 133–164. Reprinted in "The Physics of Quasicrystals" Ed. P.J. Steinhardt and Ostlund S. World Scientific Publishing Co. Singapore (1987), pp. 554–593.

[14] S. AUBRY, J.P. GOSSO, G. ABRAMOVICI, J.L. RAIMBAULT and P. QUEMERAIS, *Effective Discommensurations of the Extended Frenkel-Kontorowa Models*, To be published in Physica D.

[15] S. AUBRY, *The Devil's Staircase Transformation in Incommensurate Lattices in Seminar on the Rieman problem, Spectral Theory, Complete Integrability and Arithmetic Applications*, 1978–79 ed. D. and G. Chudnovsky Lecture Notes in Math. **925** (1982), pp. 221–0245.

[16] W. RUDIN, Real an Complex Analysis, MacGraw Hill (1979).

[17] J.N. MATHER, *"Differentiability of the Minimal Average Action as a Function of the Rotation number"*, Preprint ETH 1990.

[18] V.I. BANGERT, *"Horospheres and Peierls's Energy Barrier"* see this proceeding.

[19] A. NIVEN, *"Diophantine Approximations"*, Intersciences Publishers (1963).

[20] S. AUBRY, G. ABRAMOVICI, J.L. RAIMBAULT AND P. QUEMERAIS, *"Bipolaronic Chaotic States in the Adiabatic Holstein Model"*, In preparation.

[21] Erratum: This proof is incomplete as well as those given in ref. 6. We have to prove that $||\overline{\overline{M}}^{-1}||$ is bounded for the Sup norm (Banach norm). An erratum will be submitted soon to Physica D. The statement of theorem 3 is unchanged).

[22] A proof for the first part of this conjecture has been recently obtained (S. Aubry and R. MacKay in preparation).

HYPERSURFACES WITHOUT SELFINTERSECTIONS IN THE TORUS

VICTOR BANGERT*

Introduction. Let F be a properly embedded non-compact hypersurface in $\mathbf{R}^n$. We assume that F has no selfintersections when projected into the n-torus $T^n = \mathbf{R}^n/\mathbf{Z}^n$. This means that for every $k \in \mathbf{Z}^n$ either $T_kF \cap F = \phi$ or $T_kF = F$ where T_k denotes the translation of $\mathbf{R}^n$ by k. Our aim is to find conditions which ensure that F lies between and separates two parallel hyperplanes in $\mathbf{R}^n$, i.e. that F is similar to a leaf of a linear foliation of T^n. To this end we introduce a subgroup $G(F)$ of $\mathbf{Z}^n$ the rank of which equals the minimum of the volume growths of the two components of $\mathbf{R}^n \setminus F$. Our main result is:

THEOREM. *Let $F \subset \mathbf{R}^n$ be a properly embedded contractible C^1-hypersurface which does not have selfintersections when projected into T^n. If both components of $\mathbf{R}^n \setminus F$ have volume growth of order n then F lies between and separates two parallel hyperplanes of $\mathbf{R}^n$. If one of the components has volume growth $n-1$ then F lies between two hyperplanes which are invariant under $G(F) \subset \mathbf{Z}^n$ and which are contained in the same component of $\mathbf{R}^n \setminus F$.*

Actually we need less regularity than C^1: all we need is that $\mathbf{R}^n \setminus F$ has two components F^+ and F^- with $\partial F^+ = \partial F^- = F$ such that all homotopy groups of F^+ and F^- are trivial. This is for example true if F admits a tubular neighborhood since in this case one can apply the Seifert-van Kampen theorem, the Hurewicz theorem and the Mayer-Vietoris sequence.

The best-known example illustrating the theorem is the case when F is the lift of an orbit of a C^1-direction field on T^2, cf. [8]. Indeed, our method of proof is an extension of an idea for the case $n = 2$ which is due to A. Weil [12]. Moreover, if F is a lift to $\mathbf{R}^n$ of a leaf of a foliation on T^n defined by a closed, nowhere vanishing 1-form θ on T^n then the following simple arguments show that F lies between and separates two parallel hyperplanes. If $\pi : \mathbf{R}^n \to T^n$ denotes the projection then $\pi^*\theta = df$ and $f : \mathbf{R}^n \to \mathbf{R}$ can be written as $f(x) = Lx + g(x)$ where $L : \mathbf{R}^n \to \mathbf{R}$ is a non-zero linear form and g is $\mathbf{Z}^n$-periodic. This implies that the level surfaces of f (which are the lifts of the leaves of the foliation defined by θ) lie between and separate hyperplanes parallel to kern(L).

If the volume-growth of one of the components of $\mathbf{R}^n \setminus F$ is smaller than $n-1$ then – most likely – F need not lie between two parallel hyperplanes. It should be simple to construct such examples for the case "volume growth $< n-2$". The case "volume-growth equals $n-2$", however, is delicate and we refer to Anosov's papers [1], [2].

*Mathematisches Institut der Universität, Hebelstraße 29, D-7800 Freiburg, Germany.

The condition "no selfintersections" is natural in the context of minimal solutions of variational problems since such solutions tend to avoid unnecessary intersections, see e.g. [7]. On the other hand they need not be – at least a priori – leaves of a foliation, cf. [4]. The ideas presented here evolved in the study of minimal solutions of variational problems on T^n, cf. [10], [3], [5]. The theorem itself is used in [6], but unfortunately it is cited incorrectly there since the case that one of the components has linear volume growth is overlooked. In the last section we complete the arguments of [6] by showing that for a surface $F \subset \mathbf{R}^3$ as above no component of $\mathbf{R}^3 \setminus F$ can have linear volume growth provided that F is homotopically area-minimizing with respect to some $\mathbf{Z}^3$-periodic Riemannian metric on $\mathbf{R}^3$.

1. Notation and definitions. Translations of $\mathbf{R}^n$ by integer vectors $k \in \mathbf{Z}^n$ will be denoted by $T_k : \mathbf{R}^n \to \mathbf{R}^n$, $T_k x = x + k$.

DEFINITION 1.1. An embedded hypersurface $F \subset \mathbf{R}^n$ has no selfintersections in T^n if for every $k \in \mathbf{Z}^n$ we have either $T_k F \cap F = \phi$ or $T_k F = F$.

A submanifold F of $\mathbf{R}^n$ is proper if F is a closed subset of $\mathbf{R}^n$. As an auxiliary tool we use an arbitrary fixed scalar product on $\mathbf{R}^n$, denoted by $x \cdot y$, with associated norm $|x|$ and open balls $B(x,r)$. The scalar product determines a normalization for the Lebesgue measures vol_p on the p-dimensional affine subspaces of $\mathbf{R}^n$, $1 \leq p \leq n$. Let ω_p denote the volume of a p-dimensional unit ball.

DEFINITION 1.2. Let $A \subseteq \mathbf{R}^n$ be measurable. We say that A has (polynomial) volume growth of order p if $0 < \liminf_{r\to\infty} \left((\omega_p r^p)^{-1} \mathrm{vol}_n(A \cap B(0,r))\right) \leq \limsup_{r\to\infty} \left((\omega_p r^p)^{-1} \mathrm{vol}_n(A \cap B(0,r))\right) < \infty$. If the above limit exists we call it the (p-dimensional) growth rate $GR_p(A)$ of A.

While the property to have volume growth of order p is independent of the choice of the scalar product this is not the case for the growth rate $GR_p(A)$.

2. The group $G(F)$ and the volume growth. Suppose $F \subset \mathbf{R}^n$ is a properly embedded connected hypersurface without selfintersections in T^n. Then $\mathbf{R}^n \setminus F$ has two connected components, say F^+ and F^-, and $\partial F^+ = \partial F^- = F$. We define

$$\begin{aligned} G_+ &= G_+(F) = \{k \in \mathbf{Z}^n | T_k F^+ \subseteq F^+\} \\ G_0 &= G_0(F) = \{k \in \mathbf{Z}^n | T_k F = F\} \\ G_- &= G_-(F) = \{k \in \mathbf{Z}^n | T_k F^- \subseteq F^-\}. \end{aligned}$$

While G_0 is a subgroup of $(\mathbf{Z}^n, +)$ the sets G_+ and G_- are semigroups in $\mathbf{Z}^n$ and $-G_+ = G_-$. We intend to show that $G_+ \cup G_-$ is a subgroup. Our first observation is:

LEMMA 2.1. *$G_+ \cup G_- = \{k \in \mathbf{Z}^n | T_k F^+ \cap F^+ \neq \phi$ and $T_k F^- \cap F^- \neq \phi\}$.*

Proof. First suppose $k \in G_+ \cup G_-$, say $k \in G_+$. Then $T_k F^+ \subseteq F^+$, hence $T_k F^+ \cap F^+ \neq \phi$. Since $-G_+ = G_-$ we have $T_{-k} F^- \subseteq F^-$, hence $F^- \subseteq T_k F^-$ and thus $T_kF^- \cap F^- \neq \phi$. Now suppose $k \in \mathbb{Z}^n$ and $T_kF^+ \cap F^+ \neq \phi$ and $T_kF^- \cap F^- \neq \phi$.

(a) We first treat the case that $T_kF \cap F = \phi$. Since F is connected we have either $T_kF \subset F^+$ or $T_kF \subset F^-$, say $T_kF \subset F^+$. Then

$$F^- = (F^- \cap T_kF^+) \cup (F^- \cap T_kF^-).$$

Since F^- is connected we have either $F^- \cap T_kF^+ = \phi$ or $F^- \cap T_kF^- = \phi$. The latter alternative cannot hold by assumption. Hence $F^- \subseteq T_kF^-$. This implies $-k \in G_-$ and $k \in G_+$. Similarly the case $T_kF \subset F^-$ leads to $k \in G_-$.

(b) If $T_kF = F$ and $T_kF^+ \cap F^+ \neq \phi$ then obviously $T_kF^+ = F^+$ and $T_kF^- = F-$, i.e. $k \in G_+ \cap G_-$. □

PROPOSITION 2.2. *$G_+ \cup G_-$ is a subgroup of $\mathbb{Z}^n$.*

Proof. Since G_+ and G_- are semigroups and $-G_+ = G_-$ it suffices to show that $k + h \in G_+ \cup G_-$ if $k \in G_+$ and $h \in G_-$. Choose $x \in F^+, y \in F^-$. Since $-k \in G_-$, $-h \in G_+$ we have for all $n, m \in N$

$$x + nk - mh \in F^+$$

and

$$y - nk + mh \in F^-.$$

In particular $x - h \in F^+$ and $T_{k+h}(x - h) = x + k \in F^+$, i.e. $T_{k+h}F^+ \cap F^+ \neq \phi$. Similarly $y - k \in F^-$ and $T_{k+h}(y - k) = y + h \in F^-$, i.e. $T_{k+h}F^- \cap F^- \neq \phi$. By (2.1) we obtain $k + h \in G_+ \cup G_-$. □

Proposition 2.2 allows us to associate a subgroup $G = G(F) = G_+(F) \cup G_-(F)$ to every hypersurface F as above. Next we investigate the geometrical significance of the rank $p = p(F)$ of $G(F)$. Our first result will be that p equals the volume growth of one of the components of $\mathbb{R}^n \setminus F$ (of both if $p = n$).

Remember that for $k \in \mathbb{Z}^n \setminus G$ we have $T_kF^+ \cap F^+ = \phi$ or $T_kF^- \cap F^- = \phi$, cf. (2.1).

PROPOSITION 2.3. *Suppose $p(F) < n$. Then one of the components of $\mathbb{R}^n \setminus F$, say F^+, satisfies $T_kF^+ \cap F^+ = \phi$ for all $k \in \mathbb{Z}^n \setminus G$.*

Proof. Let $k \in \mathbb{Z}^n \setminus G$ be arbitrary and suppose that $T_kF^+ \cap F^+ = \phi$. We argue by contradiction and assume that there exists $h \in \mathbb{Z}^n \setminus G$ with $T_hF^+ \cap F^+ \neq \phi$, hence $T_hF^- \cap F^- = \phi$. Then we have

$$T_{k+h}F^+ = T_h(T_kF^+) \subseteq T_hF^- \subseteq F^+$$

and, similarly, $T_{k+h}F^- \subseteq F^-$. Hence we obtain $T_{k+h}F^+ = F^+$, $T_{k+h}F^- = F^-$ and $T_{k+h}F = F$. Since by assumption $T_{-h}F^- \cap F^- = \phi$ the equation $T_{k+h}F^- = F^-$ implies $T_kF^- = T_{-h}F^- \subseteq F^+$. Using our hypothesis $T_kF^+ \cap F^+ = \phi$ we see that $T_kF^- = F^+$ and hence $T_kF^+ = F^-$, $T_kF = F$. So we obtain $2k \in G_+ \cap G_- \subseteq G$ for every $k \in \mathbb{Z}^n \setminus G$. In particular $2\mathbb{Z}^n \subseteq G$ which contradicts our hypothesis $p(F) < n$. □

The following proposition holds also in the case $p(F) = n$ excluded in (2.3). It is a reformulation of [5], Lemma (3.1).

PROPOSITION 2.4. *Let $V \subseteq \mathbf{R}^n$ denote the linear subspace generated by G. Either we have $G = G_+ = G_-$ or there exists a unique unit vector $a \in V$ such that*

$$\{k \in G | k \cdot a > 0\} \subseteq G_+ \subseteq \{k \in G | k \cdot a \geq 0\}.$$

Remark. If $f : \mathbf{R} \to \mathbf{R}$ is a lift of an orientation preserving homeomorphism $\bar{f} : S^1 \to S^1$, i.e. f is continuous, increasing and $f(x+1) = f(x) + 1$, then every orbit $f^i(x_0) = x_i$, $i \in \mathbf{Z}$, of f gives rise to a properly embedded arc $F \subset \mathbf{R}^2$ without selfintersections in T^2 : F is the union of the line segments from (i, x_i) to $(i+1, x_{i+1})$, $i \in \mathbf{Z}$. In this case $G(F) = \mathbf{Z}^2$. If we work with the standard scalar product and choose F^+ to be the component of $\mathbf{R}^2 \setminus F$ above F then we have $a = (1+\rho^2)^{-\frac{1}{2}}(-\rho, 1)$ where ρ is the Poincaré rotation number of f. Obviously instead of F we may take every flow line of a $\mathbf{Z}^2$-periodic flow which is a suspension of f. This is closely related to A. Weil's approach [12].

Using Propositions (2.3) and (2.4) (in case $p(F) = n$) we easily obtain:

THEOREM 2.5. *Suppose $F \subset \mathbf{R}^n$ is a properly embedded connected hypersurface without selfintersections in T^n. Then $p(F)$, the rank of $G(F)$, equals the minimum of the volume growths of the components of $\mathbf{R}^n \setminus F$. We have $p(F) = n$ if and only if both components of $\mathbf{R}^n \setminus F$ have volume growth n.*

Actually we will make more precise statements about the growth rate of a component F^+ with volume growth p (this component is uniquely determined except for $p = n$). Remember that we work with a fixed arbitrary scalar product on $\mathbf{R}^n$. By vol_p we denote the Lebesgue measure on p-dimensional subspaces with the normalization induced by the given scalar product. We introduce the following notation: V is the linear subspace generated by G, $V^\perp$ its orthogonal complement and A is a bounded measurable fundamental domain for the action of G on V with $0 \in A$.

COMPLEMENT 2.6. Let F^+ be a component of $\mathbf{R}^n \setminus F$ with volume growth p. Then the growth rate $GR_p(F^+) := \lim_{r \to \infty} ((\omega_p r^p)^{-1} \mathrm{vol}_n(F^+ \cap B(0,r))$ of F^+ exists.

(a) If $G_+ = G$ and $rk(G) = n$ then

$$0 < GR_p(F^+) = \mathrm{vol}_n(A)^{-1} \mathrm{vol}_n(F^+ \cap A) < 1.$$

If $G_+ = G$ and $rk(G) = p < n$ then

$$0 < GR_p(F^+) = \mathrm{vol}_p(A)^{-1} \mathrm{vol}_n(F^+ \cap (A + V^\perp)) \leq \mathrm{vol}_p(A)^{-1} \mathrm{vol}_n([0,1]^n).$$

(b) If $G_+ \subsetneqq G$ define

$$\underline{\mu} := \inf_{k \in G} \mathrm{vol}_p(A)^{-1} \mathrm{vol}_n(F^+ \cap T_k(A + V^\perp)) \qquad \text{and}$$

$$\bar{\mu} := \sup_{k \in G} \mathrm{vol}_p(A)^{-1} \mathrm{vol}_n(F^+ \cap T_k(A + V^\perp)).$$

Then we have $GR_p(F^+) = \frac{1}{2}(\bar{\mu} + \underline{\mu})$. Moreover if $p = n$ then $0 \leq \underline{\mu} < \bar{\mu} \leq 1$, hence $0 < GR_p(F^+) < 1$, and if $p < n$ then $0 \leq \underline{\mu} < \bar{\mu} \leq \mathrm{vol}_p(A)^{-1} \mathrm{vol}_n([0,1]^n)$, hence $0 < GR_p(F^+) < \mathrm{vol}_p(A)^{-1} \mathrm{vol}_n([0,1]^n)$.

Proof of (2.6) (a). We first treat the case $p = rk(G) = n$. Since $T_k F^+ = F^+$ for all $k \in G$ we can represent F^+ as the disjoint union

$$F^+ = \bigcup_{k \in G} T_k(F^+ \cap A).$$

Then

$$\bigcup_{\substack{k \in G \\ |k| \le r-D}} T_k(F^+ \cap A) \subseteq F^+ \cap B(0,r) \subseteq \bigcup_{\substack{k \in G \\ |k| \le r+D}} T_k(F^+ \cap A) \tag{2.7}$$

where D denotes the diameter of A. On the other hand there exists $C > 0$ such that for all $r > 0$:

$$\left| \#\{k \in G \mid |k| \le r\} \operatorname{vol}_n(A) - \omega_n r^n \right| \le C(r^{n-1} + 1) \tag{2.8}$$

From (2.7) and (2.8) we conclude

$$\lim_{r \to \infty} \left((\omega_n r^n)^{-1} \operatorname{vol}_n(F^+ \cap B(0,r)) \right) = \operatorname{vol}_n(A)^{-1} \operatorname{vol}_n(F^+ \cap A). \tag{2.9}$$

Since both F^+ and F^- are non-empty, open and G-invariant we have

$$0 < \operatorname{vol}_n(A)^{-1} \operatorname{vol}_n(F^+ \cap A) < 1.$$

The proof for $p < n$ is similar: Following (2.3) we choose F^+ to be the component of $\mathbf{R}^n \setminus F$ satisfying $T_k F^+ \cap F^+ = \phi$ for all $k \in \mathbf{Z}^n \setminus G$. Then we have

$$T_k \left(F^+ \cap (A + V^\perp)\right) \cap \left(F^+ \cap (A + V^\perp)\right) = \phi$$

for all $k \in \mathbf{Z}^n \setminus \{0\}$ and $\bigcup_{k \in G} T_k \left(F^+ \cap (A + V^\perp)\right) = F^+$. This implies

$$0 < \operatorname{vol}_n \left(F^+ \cap (A + V^\perp)\right) \le \operatorname{vol}_n([0,1]^n) \tag{2.10}$$

and

$$F^+ \cap B(0,r) \subseteq \bigcup_{\substack{k \in G \\ |k| \le r+D}} T_k \left(F^+ \cap (A + V^\perp)\right) \tag{2.11}$$

where, again, $D = \operatorname{diam}(A)$. Using (2.8) with n replaced by p we obtain from (2.11)

$$\limsup_{r \to \infty} \left((\omega_p r^p)^{-1} \operatorname{vol}_n(F^+ \cap B(0,r)) \right) \le \operatorname{vol}_p(A)^{-1} \operatorname{vol}_n \left(F^+ \cap (A + V^\perp)\right). \tag{2.12}$$

To prove a reversed inequality we note that for every $\varepsilon > 0$ there exists $R > 0$ such that

$$\operatorname{vol}_n \left(F^+ \cap (A + V^\perp) \cap B(0,R)\right) \ge \operatorname{vol}_n \left(F^+ \cap (A + V^\perp)\right) - \varepsilon.$$

Since

$$\bigcup_{\substack{k\in G\\ |k|\le r-R}} T_k\left(F^+\cap(A+V^\perp)\cap B(0,R)\right)\subseteq F^+\cap B(0,r)$$

we obtain

(2.13)

$$\mathrm{vol}_p(A)^{-1}\left(\mathrm{vol}_n(F^+\cap(A+V^\perp))-\varepsilon\right)\le\liminf_{r\to\infty}\left((\omega_p r^p)^{-1}\,\mathrm{vol}_n(F^+\cap B(0,r))\right).$$

This implies that the lim sup in (2.12) and the lim inf in (2.13) coincide and equal $\mathrm{vol}_p(A)^{-1}\,\mathrm{vol}_n\left(F^+\cap(A+V^\perp)\right)$. Now (2.10) completes the proof of (2.6) (a).

Proof of (2.6) (b): First we prove the estimate on $\bar{\mu}$. The case $p=n$, i.e. $V^\perp=\{0\}$, is trivial. If $p<n$ we have $T_kF^+\cap F^+=\phi$ for all $k\in\mathbb{Z}^n\setminus G$, cf. (2.3). This implies that $X:=\bigcup_{k\in G}\left(T_kF^+\cap(A+V^\perp)\right)$ satisfies $T_kX\cap X=\phi$ for all $k\in\mathbb{Z}^n\setminus\{0\}$. Hence $\mathrm{vol}_n(X)\le\mathrm{vol}_n([0,1]^n)$ and this proves our estimate on $\bar{\mu}$. To show that $\underline{\mu}<\bar{\mu}$ let a denote the unit vector in V such that $T_kF^+\subseteq F^+$ if $k\in G$ and $k\cdot a>0$, cf. (2.4). If $k\in G$ and $k\cdot a>0$ then $k\notin G_-$ so that we actually have $T_kF^+\subsetneqq F^+$. Hence

$$\mathrm{vol}_n\left(F^+\cap T_{j-k}(A+V^\perp)\right)<\mathrm{vol}_n\left(F^+\cap T_j(A+V^\perp)\right)$$

for some $j\in G$ and this implies $\underline{\mu}<\bar{\mu}$. Similarly we have the following important fact: if $k,j\in G$ and $k\cdot a>j\cdot a$ then

$$\mathrm{vol}_n\left(F^+\cap T_k(A+V^\perp)\right)\ge\mathrm{vol}_n\left(F^+\cap T_j(A+V^\perp)\right).$$

Now we prove that the growth rate $GR_p(F^+)$ of F^+ equals $\frac{1}{2}(\bar{\mu}+\underline{\mu})$. The preceding inequality implies that for every $\varepsilon>0$ there exists $M>0$ such that

$$\bar{\mu}-\varepsilon\le\mathrm{vol}_p(A)^{-1}\,\mathrm{vol}_n\left(F^+\cap T_k(A+V^\perp)\right)$$

if $k\in G$ and $k\cdot a\ge M$ and such that

$$\mathrm{vol}_p(A)^{-1}\,\mathrm{vol}_n\left(F^+\cap T_k(A+V^\perp)\right)\le\underline{\mu}+\varepsilon$$

if $k\in G$ and $k\cdot a\le -M$. In view of the inclusion

$$F^+\cap B(0,r)\subseteq\bigcup_{\substack{k\in G\\ |k|\le r+D}}\left(F^+\cap T_k(A+V^\perp)\right)$$

we subdivide the set of $k\in G$ with $|k|\le r+D$ into those with $k\cdot a\ge M$, those with $|k\cdot a|<M$ and those with $k\cdot a\le -M$. The first and the third set have the same number of elements and the number of elements of the second set grows like r^{p-1}. Using (2.8) with n replaced by p we obtain

(2.14) $$\limsup_{r\to\infty}\left((\omega_p r^p)^{-1}\,\mathrm{vol}_n\left(F^+\cap B(0,r)\right)\right)\le\frac{1}{2}(\bar{\mu}+\underline{\mu}+\varepsilon).$$

Since $X=\bigcup_{k\in G}\left(T_kF^+\cap(A+V^\perp)\right)$ has finite volume there exists $R>0$ such that for all $k\in G$ we have

$$\mathrm{vol}_n\left(F^+\cap T_k\left((A+V^\perp)\cap B(0,R)\right)\right)\ge\mathrm{vol}_n\left(F^+\cap T_k(A+V^\perp)\right)-\varepsilon.$$

Now similar arguments as above show

(2.15) $$\frac{1}{2}(\bar{\mu}-\varepsilon+\underline{\mu})-\varepsilon\,\mathrm{vol}_p(A)^{-1}\le\liminf_{r\to\infty}\left((\omega_p r^p)^{-1}\,\mathrm{vol}_n\left(F^+\cap B(0,r)\right)\right)$$

Complement (2.6) (b) follows from (2.14) and (2.15).

3. Existence of linear bounds in the contractible case. In this section we prove the theorem stated in the introduction.

THEOREM 3.1. *Let $F \subset \mathbb{R}^n$ be a properly embedded contractible C^1-hypersurface without selfintersections in T^n. If $p = rk\,(G(F)) = n$ then F lies between and separates two parallel hyperplanes of $\mathbb{R}^n$.*

Remarks. 1. As a consequence we obtain $G = \mathbb{Z}^n$ and $G_+ \subsetneqq G$, cf. Lemma (2.1). The unit vector a with

$$\{k \in \mathbb{Z}^n | k \cdot a > 0\} \subseteq G_+(F) \subseteq \{k \in \mathbb{Z}^n | k \cdot a \geq 0\}$$

(see (2.4)) is orthogonal to the hyperplanes mentioned above.

2. It seems likely that the condition "F contractible" can be relaxed, say to "all homotopy groups of F are finitely generated". However some condition is necessary, cf. example 5 below.

Proof. First we note that all homotopy groups of F^+ and F^- vanish. This follows from the Seifert-van Kampen theorem in dimension one and from the Mayer-Vietoris sequence and the Hurewicz theorem in higher dimensions. We choose a basis $\{k_1, \ldots, k_n\} \subseteq G_+$ of G and denote by P the parallelotope $P = \left\{ \sum_{i=1}^n s_i k_i | 0 \leq s_i \leq 1 \right\}$. We set

$$H = \bigcup \{T_k(P) | k \in G_+\}.$$

If $G_+ \subsetneqq G$ then Proposition (2.4) implies that H is contained in some halfspace and contains some halfspace of $\mathbb{R}^n$. If $G_+ = G$ then $H = \mathbb{R}^n$, but we will show that this case cannot occur.

Our aim is to construct a continuous map $f : H \to F^+$ such that

$$f(x + k) = f(x) + k \quad \text{for } x \in H, k \in G_+. \tag{3.2}$$

We first choose on arbitrary point $f(0) \in F^+$. Then (3.2) forces us to define $f\left(\sum_{i=1}^n \varepsilon_i k_i\right) = f(0) + \sum_{i=1}^n \varepsilon_i k_i$ for every vertex $\sum_{i=1}^n \varepsilon_i k_i$ of $P, \varepsilon_i \in \{0,1\}$. Since $\sum_{i=1}^n \varepsilon_i k_i \in G_+$ and $f(0) \in F^+$ we have indeed $f\left(\sum_{i=1}^n \varepsilon_i k_i\right) \in G_+$. By induction we may assume that $f : P^{(j-1)} \to F^+$ has been defined on the $(j-1)$ skeleton $P^{(j-1)}$ of $P, 0 < j \leq n$, such that (3.2) is satisfied. For a sequence $1 \leq i_1 < \cdots < i_j \leq n$ let

$$G_{i_1 \ldots i_j} = \left\{ \sum_{m=1}^j s_m k_{i_m} | 0 \leq s_m \leq 1 \right\}$$

be a j-face of P containing $0 \in \mathbb{R}^n$. Then $f | \partial G_{i_1 \ldots i_j}$ is already defined and satisfies (3.2). Since F^+ is contractible we can extend $f | \partial G_{i_1 \ldots i_j}$ to a map $f : G_{i_1 \ldots i_j} \to F^+$. Thus we extend f to all the j-faces of P containing $0 \in \mathbb{R}^n$. Now an arbitrary j-face of P is of the form $T_k(G_{i_1\dots i_j})$ for some $k = \sum_{i=1}^{n} \varepsilon_i k_i \in G_+ \setminus \{0\}$, $\varepsilon_i \in \{0,1\}$, $\varepsilon_i = 0$ if $i \in \{i_1, \dots, i_j\}$. So the extension of f to $T_k(G_{i_1\dots i_j})$ is determined by (3.2). Since $f|P^{(j-1)}$ satisfies (3.2) this is indeed an extension of $f|P^{(j-1)}$ and since $k \in G_+$ and $f(G_{i_1\dots i_j}) \subseteq F^+$ we have $f\left(T_k(G_{i_1\dots i_j})\right) \subseteq F^+$. This completes the induction so that we now have $f : P \to F^+$ satisfying (3.2). Again (3.2) determines how we have to extend f from P to all of H. That this extension is indeed possible and maps H into F^+ follows as above.

The first consequence of the existence of f is that $G_+ \subsetneqq G$. Otherwise we have $H = \mathbb{R}^n$ so that $f : \mathbb{R}^n \to F^+$ is a continuous map satisfying $f(x+k) = f(x)+k$ for the lattice $G \subset \mathbb{R}^n$. This implies that $|f(x) - x|$ is uniformly bounded. Hence f can be continuously extended to $S^n = \mathbb{R}^n \cup \{\infty\}$ by $f(\infty) := \infty$ and f is homotopic to id_{S^n}. Thus $f : S^n \to S^n$ has degree 1 and this implies $f(\mathbb{R}^n) = \mathbb{R}^n$ in contradiction to $f(\mathbb{R}^n) \subseteq F^+ \subsetneqq \mathbb{R}^n$. So we have $G_+ \subsetneqq G$. According to Proposition (2.4) there exists a unit vector $a \in \mathbb{R}^n$ such that

$$\{k \in G | k \cdot a > 0\} \subseteq G_+ \subseteq \{k \in G | k \cdot a \geq 0\}.$$

Now the existence of $f : H \to F^+$ satisfying (3.2) allows us to prove that $\sup_{x \in F}(x \cdot a) < \infty$, i.e. that F lies below some hyperplane with normal a. We first show that we can find balls of arbitrarily large radius inside F^+. As above there exists $C > 0$ such that $|f(x) - x| < C$ for all $x \in H$. For a given radius R choose a ball B_1 of radius $R + C$ inside H and let B_0 denote the ball of radius R with the same center. Then the affine homotopy from the inclusion $i_{\partial B_1} : \partial B_1 \to \mathbb{R}^n$ to $f|\partial B_1$ does not intersect B_0. This implies $f(B_1) \supseteq B_0$: if there would exist $p \in B_0 \setminus f(B_1)$ we could contract $i_{\partial B_1}$ to a point in $\mathbb{R}^n \setminus \{p\}$. Since $f(B_1) \subseteq F^+$ we have found a ball B_0 of radius R in F^+. In particular F^+ contains some fundamental domain $\tilde{P}$ of $\mathbb{R}^n/G$. We are going to prove

$$\sup_{x \in F}(x \cdot a) \leq \sup_{x \in \tilde{P}}(x \cdot a) < \infty$$

which shows that F lies below some hyperplane with normal a: if $y \in \mathbb{R}^n$ and $y \cdot a > \sup_{x \in \tilde{P}}(x \cdot a)$ then $y = x + k$ for some $x \in \tilde{P}$ and some $k \in G$ with $k \cdot a > 0$. Since $\tilde{P} \subset F^+$ and $k \in G_+$ we have $y \in F^+$, i.e. $y \notin F$. Analogous arguments show that $\inf_{x \in F}(x \cdot a) > -\infty$ and this completes the proof of Theorem (3.1). □

Before we treat the case $p = n - 1$ we formulate a simple lemma.

LEMMA 3.3. *Let $f_0 : M \to N$ be a properly embedded connected hypersurface and let f_1 be homotopic to f_0 by a proper homotopy $f : M \times [0,1] \to N$. Suppose x and y are points in N which lie in different components of $N \setminus f_0(M)$. If $\{x, y\} \cap f(M \times [0,1]) = \phi$ then x and y lie in different components of $N \setminus f_1(M)$.*

Proof. If $\gamma : [0,1] \to N$ is a path from x to y and if $g : M \to N \setminus \{x, y\}$ is proper then the intersection number $\#(\gamma, g) \mod 2$ is defined and invariant under proper homotopies of g in $N \setminus \{x, y\}$, cf. [9], Chapter 5. Since $\#(\gamma, f_0) = 1 \mod 2$ by assumption we have $\#(\gamma, f_1) = 1 \mod 2$ for every path γ from x to y. Hence x and y lie in different components of $N \setminus f_1(M)$. □

THEOREM 3.4. *Let $F \subset \mathbb{R}^n$ be a properly embedded contractible C^1-hypersurface without selfintersections in T^n. If $p = rk(G(F)) = n-1$ then F lies between two hyperplanes which are invariant under $G(F)$. These hyperplanes lie in the same component of $\mathbb{R}^n \setminus F$ and $G_+ \subsetneqq G$.*

Proof. According to Proposition (2.3) one of the components of $\mathbb{R}^n \setminus F$, say F^+, satisfies $T_k F^+ \cap F^+ = \phi$ for all $k \in \mathbb{Z}^n \setminus G$. Choose a basis $\{k_1, \dots, k_{n-1}\} \subset G_+$ of G and let P denote the $(n-1)$-dimensional parallelotope generated by $k_1, \dots k_{n-1}$. As in the proof of Theorem (3.1) we can find a continuous map $f : \bigcup_{k \in G_+} T_k P \to F^+$ such that $f(x+k) = f(x)+k$ for all $k \in G_+$. First we prove that $G_+ \subsetneqq G$. Otherwise $\bigcup_{k \in G_+} T_k P =: H$ is a hyperplane in $\mathbb{R}^n$. Due to the periodicity $f(x+k) = f(x)+k$ there exists $C > 0$ such that $|f(x) - x| < C$ for all $x \in H$. Now we choose an arbitrary point $x_0 \in F^+$ and an integer vector $k \in \mathbb{Z}^n \setminus G$ such that $x_1 = x_0 + k$ and $y_1 = x_0 - k$ lie in different components of $\mathbb{R}^n \setminus H$ and have distance $> C$ from H. Since f and the inclusion $i_0 : H \to \mathbb{R}^n$ are properly homotopic by an affine homotopy in $\mathbb{R}^n \setminus \{x_1, y_1\}$ we conclude from Lemma (3.3) that x_1 and y_1 lie in different components of $\mathbb{R}^n \setminus f(H)$. On the other hand $x_1 = T_k x_0 \in F^-$ and $y_1 = T_{-k} x_0 \in F^-$ since $x_0 \in F^+$ and $k \in \mathbb{Z}^n \setminus G$. The preceding two statements contradict the facts that $f(H) \subset F^+$ and that F^- is connected. This proves $G_+ \subsetneqq G$. The proof of the remaining statements is similar: We consider the hyperplane $H := \bigcup_{k \in G} T_k(P)$. For $j \in \mathbb{Z}^n \setminus G$ we define $f_j : H \to \mathbb{R}^n$ by

$$f_j(x+k) = f(x) + k + j$$

if $x \in P$ and $k \in G$. This is well-defined since $f(x+k) = f(x) + k$ for all $x \in P$, $k \in G_+$. Since $k + j \in \mathbb{Z}^n \setminus G$ and since $f(P) \subset F^+$ we have $f_j(H) \subset F^-$. As above there exists $C > 0$ such that $|f_j(x) - i_j(x)| < C$ for all $x \in H$ where $i_j : H \to \mathbb{R}^n$ is defined by $i_j(x) = x + j$. From Lemma (3.3) we conclude that $f_j(H)$ separates two hyperplanes H_j^0 and H_j^1 parallel to H and at distance C from each other. Since $f_j(H) \subset F^-$ and F^+ is connected F^+ cannot intersect both H_j^0 and H_j^1. So, for every $j \in \mathbb{Z}^n \setminus G$ we obtain a hyperplane $H_j^{\varepsilon_j} \subset F^-$ parallel to H and at distance $\leq C$ from $i_j(H) = T_j(H)$. Obviously this implies the existence of two hyperplanes $H_0 \subset F^-$ and $H_1 \subset F^-$ such that F^+ lies between H_0 and H_1. □

Examples. We first give a series of "trivial" examples of properly embedded contractible hypersurfaces $F \subset \mathbb{R}^n$ without selfintersections in T^n which show that $p = rk(G(F))$ can be every integer in $\{0, \dots, n\}$.

1. If F is an affine hyperplane then $G(F) = \mathbb{Z}^n$.
2. The curve $F \subset \mathbb{R}^2$ pictured below has $G(F) = \mathbb{Z} \cdot e_1$, i.e. $p = 1$.
3. The curve $F \subset \mathbb{R}^2$ pictured below has $G(F) = \{0\}$, i.e. $p = 0$.
4. Examples 2 and 3 obviously generalize to hypersurfaces $F \subset \mathbb{R}^n$, $n > 2$, (with the properties stated above) such that $p = n-1$ resp. $p = 0$. To obtain $F \subset \mathbb{R}^n$ with $p \in \{1, \dots, n-2\}$, choose $F' \subset \mathbb{R}^{n-p}$ with $G(F') = \{0\}$ and set $F := F' \times \mathbb{R}^p \subset \mathbb{R}^n$.

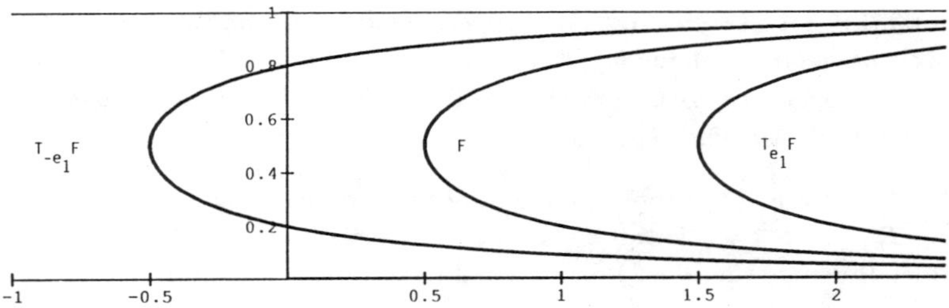

Figure 1

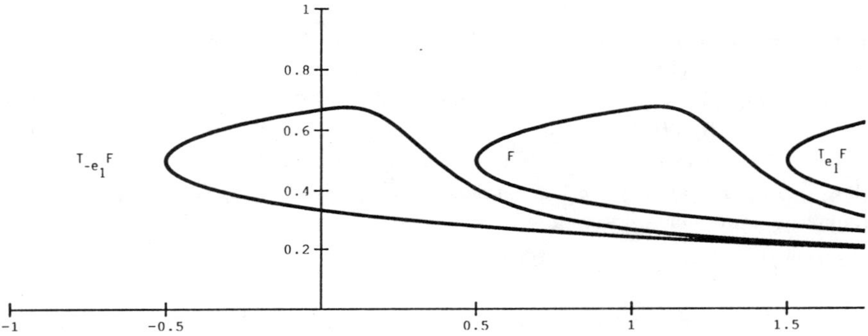

Figure 2

Next we describe examples which show that some condition on the topology of F is necessary in Theorems (3.1) and (3.4).

5. Let A be the union of all translates of the coordinate axes in $\mathbb{R}^3$ by integer vectors, $A = (\mathbb{R} \times \mathbb{Z} \times \mathbb{Z}) \cup (\mathbb{Z} \times \mathbb{R} \times \mathbb{Z}) \cup (\mathbb{Z} \times \mathbb{Z} \times \mathbb{R})$. We smoothe the boundary of a small tubular neighborhood of A in a $\mathbb{Z}^3$-periodic way and obtain a properly embedded surface F such that $T_k F = F$ for all $k \in \mathbb{Z}^3$, in particular $G(F) = \mathbb{Z}^3$. Obviously F is not contained between two parallel hyperplanes.
6. We let $\tilde{F}$ be the boundary of a small tubular neighborhood of $(\mathbb{R} \times \mathbb{Z} \times \{0\}) \cup (\mathbb{Z} \times \mathbb{R} \times \{0\})$ smoothed in a $\mathbb{Z}^2 \times \{0\}$-periodic way. Let $\tilde{F}^-$ denote the component of $\mathbb{R}^3 \setminus \tilde{F}$ containing the line $\{\frac{1}{2}\} \times \{\frac{1}{2}\} \times \mathbb{R}$. Now we remove a small disk from $\tilde{F}$ and glue an embedded proper tube homeomorphic to $S^1 \times [0, \infty)$ into the boundary. This tube shall approach the halfline $\{\frac{1}{2}\} \times \{\frac{1}{2}\} \times [0, \infty)$ so fast that it does not have selfintersections in T^3,

cf. example 3 above. Now we extend this construction $\mathbf{Z}^2 \times \{0\}$-periodically to obtain a properly embedded surface $F \subset \mathbf{R}^3$ without selfintersections in T^3. We have $G(F) = \mathbf{Z}^2 \times \{0\}$ but F is not contained between two parallel hyperplanes.

Finally we approach the problem to construct hypersurfaces $F \subset \mathbf{R}^n$ which show that the hypothesis $p \geq n-1$ in (3.1) and (3.4) cannot be omitted. For $n \geq 3$ it should be simple to construct hypersurfaces $F \subset \mathbf{R}^n$ as above with $p = 0$ which are not contained between two parallel hyperplanes: they can be obtained as tubes with rapidly decreasing radius about appropriate curves $\gamma : [0, \infty) \to \mathbf{R}^n$. An explicit construction may be cumbersome and so we omit it. Repeating the product construction from example 4 one can then obtain such examples with $p \leq n-3$. The case $p = n-2$ is delicate, already in the case $n = 2$. In [2] Anosov presents the very subtle construction of a proper curve $\gamma : [0, \infty) \to \mathbf{R}^2$ without selfintersections on T^2 which is not contained in any strip between two parallel lines. This makes it plausible that examples as above will also exist in the case $p = n-2$.

4. Least area surfaces without selfintersections in the 3-torus. The purpose of this section is to prove Theorem (4.1) below, a result which is used in [6], Theorem 3.4. As a side-effect this section shows how hard it is in general to control the shape of non-compact surfaces in $\mathbf{R}^3$. Thus it illustrates the special character of the theorems in the preceding section.

We start with a short review of the notions used in [6]. Let g be a $\mathbf{Z}^3$-periodic Riemannian metric on $\mathbf{R}^3$. A surface $F \subset \mathbf{R}^3$ is called homotopically g-area minimizing if every compactly supported homotopy of F does not decrease the g-area of the deformed part of F. Following [6] we let $\mathfrak{F} = \mathfrak{F}(g)$ denote the set of all properly embedded submanifolds $F \subset \mathbf{R}^3$ with the following properties

(F_1) F is homeomorphis to the plane

(F_2) F is homotopically g-area minimizing

(F_3) F does not have selfintersections in $T^3 = \mathbf{R}^3/\mathbf{Z}^3$.

We are concerned with the first claim in [6], Theorem 4.3: every $F \in \mathfrak{F}$ lies between two parallel affine planes. Theorems (2.5), (3.1) and (3.4) show that to prove this claim it suffices to prove that both components of $\mathbf{R}^3 \setminus F$ have at least quadratic volume growth. In [6], Corollary (3.3), it is proved that both components of $\mathbf{R}^3 \setminus F$ have infinite volume. By an error in counting the author stated in the proof of [6], Theorem (3.4), that this suffices to show that F lies between two parallel planes. Instead it remains to prove:

THEOREM 4.1. *For every $F \in \mathfrak{F}(g)$ no component of $\mathbf{R}^3 \setminus F$ has linear volume growth.*

In spite of the strong assumption (F_2) which implies in particular that the principal curvatures of F are uniformly bounded the proof of (4.1) is complicated. If we would assume that F is absolutely g-area minimizing it would be much simpler.

Since g and the euclidean metric are uniformly comparable there exists $Q \geq 1$ such that every $F \in \mathfrak{F}(g)$ has the following quasiminimality property with respect to the area A induced from the euclidean metric:

(4.2) If $K \subset F$ is compact and if $i_t, 0 \leq t \leq 1$, is a homotopy of the inclusion $i_0 : F \to \mathbf{R}^3$ with support in K then $A(K) \leq QA(i_1|K)$.

Actually we will only use this quasiminimality of F and so the metric g itself will not be used in the sequel: consequently all metric concepts like balls $B(p,r)$ and distance d in $\mathbf{R}^3$, unit normal N of F, principal curvatures, inner metric d_F with balls $B_F(p,r) \subset F$ on F will be induced by the euclidean metric.

Next we define some concepts which will be important in the proof. We assume that $F \in \mathfrak{F}$ and that the component F^+ of $\mathbf{R}^3 \setminus F$ has linear volume growth. An important consequence of this assumption is the fact that F^+ has large thin parts. This property will be formalized as follows. Let N denote the unit normal of F pointing into F^+. We define $f : F \to \mathbf{R}$ by

$$f(p) := \sup\{t | d(p + tN(p), F) = t\}$$

where $d(q, F)$ denotes the distance from q to F. Note that $f(p) < \infty$ since otherwise F^+ would contain a halfspace. Later we will prove that f attains arbitrarily small positive values. According to [11], Theorem 3, the principal curvatures of F are uniformly bounded. So elementary differential geometry shows that there exists $t_0 > 0$ such that the following is true on the open subset $U = \{p \in F | f(p) < t_0\}$ of F: there exists an involution $i : U \to U$ without fixed points such that $f = f \circ i$ and

$$p + f(p)N(p) = i(p) + f(i(p))N(i(p))$$

for all $p \in U$. The involution i and the function $f|U$ are smooth. Since F is embedded with bounded principal curvatures we can make $N + N \circ i$ small uniformly on U if we choose t_0 sufficiently small. Then $i : U \to U$ will almost be an isometry. To be specific we choose t_0 so small that $i : U \to U$ is locally Lipschitz with constant 2.

Finally, for every $\varepsilon > 0$ we let F_ε^+ denote the set of points $p \in F^+$ such that the volume of the component of p in $F^+ \cap B(p, 2)$ is smaller than ε. Then we set

$$F_\varepsilon := F \cap \text{ closure } (F_\varepsilon^+)$$

Since the ball of radius $f(p)$ about $p + f(p)N(p)$ is contained in F^+ we have:

(4.3) If $\varepsilon < \frac{4\pi}{3}$ and if $p \in F$ has surface distance $d_F(p, F_\varepsilon) < 1$ from F_ε then $\frac{4\varepsilon}{3} f(p)^3 \leq \varepsilon$. In particular, if $q \in F_\varepsilon$ and $\varepsilon \leq \frac{4\pi}{3} t_0^3$ then $B_F(q, 1) \subset U$.

Now we will give an outline of the proof of (4.1). We want to derive a contradiction from the assumption that there exists $F \in \mathfrak{F}$ such that F^+ has linear volume growth. For every $\delta > 0$ we will construct a disk $D \subset U$ with large area such that

$f(p) < \delta$ for all $p \in D$. This disk will be long and thin: it will be part of a tubular neighborhood of a long curve in F_ε. However, it will satisfy

$$L(\partial D) \leq CA(D) \tag{4.4}$$

for some constant C which depends only on F. Then

$$Z : \partial D \times [0,1] \to \mathbf{R}^3, \quad Z(p,t) = (1-t)p + ti(p)$$

parametrizes a cylinder with area $A(Z) \leq 4\delta L(\partial D)$ which joins ∂D to $\partial(iD)$. We will show that D and iD are disjoint. Then, replacing $D \cup iD$ by Z we could decrease the g-area of F if δ is chosen small enough. This would be a contradiction if F were even "absolute g-area minimizing". Unfortunately the hypothesis "homotopically g-area minimizing" does not apply in this case since $D \cup iD$ and Z are topologically different. So the following type of "surgery" will be used. We choose an arc Γ in $F \setminus (D \cup iD)$ joining D and iD, an arc Γ' on Z joining the endpoints of Γ and a homotopy H from Γ to Γ'. Then (the inclusion into $\mathbf{R}^3$ of) the topological disk on F which consists of the union of D, iD and a small tubular neighborhood of Γ is homotopic with fixed boundary to a map into $\mathbf{R}^3$ with area approximately $A(Z) + 2A(H)$. This map joins two copies of H along Γ' to a map of the disk to Z

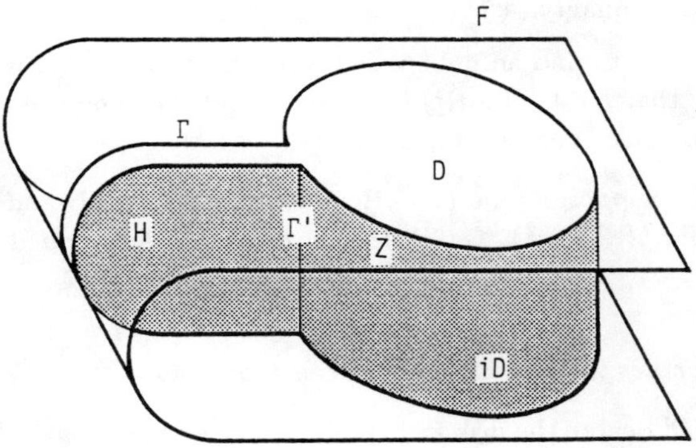

Figure 3

According to (4.2) our assumption that F be homotopically minimizing implies

$$A(D) + A(iD) \leq Q\,(A(Z) + 2A(H))\,. \tag{4.5}$$

Combining (4.4), (4.5) with the inequality $A(Z) \leq 4\delta L(\partial D)$ we obtain

$$A(D) \leq Q\,(4\delta C A(D) + 2A(H))\,.$$

Now the main point of the proof is to show that we can perform this construction so that $A(H)$ remains bounded independently of the choice of δ and $A(D)$. Then, choosing δ small and $A(D)$ large we will obtain a contradiction to the preceding inequality.

Next we present the details of the proof of (4.1). We assume that F^+ has linear volume growth. So Theorem (2.5) implies that $rk(G(F)) = 1$. Changing coordinates by an element of $SL(n, \mathbf{Z})$ we may assume that $G(F)$ is generated by me_1, for some $m \in \mathbf{N}$, and that $me_1 \in G_+(F)$. We use the standard scalar product on $\mathbf{R}^3$ and we group the components of $p \in \mathbf{R}^3$ as $p = (x, y) \in \mathbf{R} \times \mathbf{R}^2$.

LEMMA 4.6. *For every $\varepsilon > 0$ there exists $R > 0$ such that every point $(x, y) \in F$ with $|y| > R$ lies in F_ε.*

Proof. According to (2.3) we have $T_k F^+ \cap F^+ = \phi$ if $k \in \{0\} \times \mathbf{Z}^2$. This implies:

$$\text{vol}_3 \left(F^+ \cap ([a, b] \times \mathbf{R}^2)\right) \leq b - a. \tag{4.7}$$

Now assume that our claim is not true. Then we can find a sequence $p_i = (x_i, y_i) \in F^+$ with $|y_i| \to \infty$, $|y_i - y_j| > 4$ for $i \neq j$ and $p_i \notin F_\varepsilon^+$. Since $T_{me_1} F^+ \subseteq F^+$ implies $T_{me_1}(F^+ \setminus F_\varepsilon^+) \subseteq (F^+ \setminus F_\varepsilon^+)$ we can replace p_i by $T_{l_i m_i e_1} p_i = (x_i + l_i m_i, y_i)$, $l_i \in \mathbf{N}$, without changing the above-mentioned properties of the p_i. Choose $i_0 > (m+4)\varepsilon^{-1}$. We may assume that there exists $i_1 \in \{1, \ldots i_o\}$ such that $|x_i - x_{i_1}| \leq m/2$ for all $i \in \{1, \ldots, i_o\}$. Since $|y_i - y_j| > 4$ for $i \neq j$, the balls $B(p_i, 2)$, $1 \leq i \leq i_0$, are pairwise disjoint. Moreover we have $B(p_i, 2) \subset \left(\left[x_{i_1} - \frac{m}{2} - 2, x_{i_1} + \frac{m}{2} + 2\right] \times \mathbf{R}^2\right)$. Hence our assumption $p_i \notin F_\varepsilon^+$ implies that $\text{vol}_3 \left(\bigcup_{i=1}^{i_0} B(p_i, 2) \cap F^+\right) \geq i_0 \varepsilon > m+4$. This contradicts inequality (4.7). □

Our next aim is to find an unbounded component of F_ε. According to (4.6) it suffices to show that $F \cap (\{x\} \times \mathbf{R}^2)$ has an unbounded component for some $x \in \mathbf{R}$. First we treat bounded components of $F \cap (\{x\} \times \mathbf{R}^2)$.

LEMMA 4.8. *Suppose F and $\{x\} \times \mathbf{R}^2$ intersect transversely and Γ is a bounded component of $F \cap (\{x\} \times \mathbf{R}^2)$. Let D denote the disk in F bounded by Γ. Then we have*

$$D \subset ([x - Q, x + Q] \times \mathbf{R}^2)$$

where $Q \geq 1$ denotes the quasiminimality constant from (4.2).

Proof. Let D' denote the disk in $\{x\} \times \mathbf{R}^2$ bounded by Γ. Since $T_k \Gamma \cap \Gamma = \phi$ for all $k \in \{0\} \times \mathbf{Z}^2$ we have $A(D') \leq 1$. Now the quasiminimality (4.2) of F implies $A(D) \leq Q$. On the other hand [6], Lemma (2.4), says that the area of every metric ball $B_F(p, r)$ on F satisfies

$$A\left(B_F(p, r)\right) \geq (\pi/Q) r^2$$

Hence, if $p \in D$ and $d_F(p, \Gamma) = r$ then

$$(\pi/Q) r^2 \leq A(D) \leq Q,$$

hence $r < Q$. This implies our claim. □

LEMMA 4.9. *There exists $x \in \mathbf{R}$ such that F and $\{x\} \times \mathbf{R}^2$ intersect transversely and such that $F \cap (\{x\} \times \mathbf{R}^2)$ has an unbounded component.*

Proof. Since a 1-dimensional proper submanifold of a plane has an unbounded component if (and only if) its complement has at least two unbounded components if suffices to prove that $F^+ \cap (\{x \times \mathbf{R}^2)$ and $F^- \cap (\{x\} \times \mathbf{R}^2)$ have unbounded components. We start by constructing a proper curve $\gamma : [0, \infty) \to F^+$ such that $\gamma(j) = \gamma(0) + jme_1$ for all $j \in \mathbf{N}$. If $\gamma(0)$ is chosen arbitrarily we can find $\gamma : [0,1] \to F^+$ joining $\gamma(0)$ to $\gamma(1) = \gamma(0) + me_1$ since F^+ is connected. Then we define $\gamma(t)$ for $t \in [j, j+1], j \in \mathbf{N}$, by

$$\gamma(t) = \gamma(t-j) + jme_1.$$

Now we choose $x \in \mathbf{R}$ such that F and $\{x\} \times \mathbf{R}^2$ intersect transversely and such that $x > x(0) + Q$ where $\gamma(0) = (x(0), y(0))$. Then the intersection number of γ with $\{x\} \times \mathbf{R}^2$ is defined and

$$\#(\gamma, \{x\} \times \mathbf{R}^2) \equiv 1 \mod 2.$$

Assuming that all components of $F^+ \cap (\{x\} \times \mathbf{R}^2)$ are bounded one easily finds a disk $D' \subset (\{x\} \times \mathbf{R}^2)$ with $\partial D' =: \Gamma \subset F$ and such that $\#(\gamma, D') \equiv 1 \mod 2$. Let then D denote the disk on F bounded by Γ. According to (4.8) the disks D and D' are rel(Γ)-homotopic in $[x-Q, x+Q] \times \mathbf{R}^2$. Since $\gamma(0) \notin [x-Q, x+Q] \times \mathbf{R}^2$ we have $\#(\gamma, D) \equiv \#(\gamma, D') \equiv 1 \mod 2$ and this contradicts the fact that $\gamma([0,\infty)) \subset F^+$ is disjoint from $D \subset F$. Hence $F^+ \cap (\{x\} \times \mathbf{R}^2)$ has an unbounded component. To show that $F^- \cap (\{x\} \times \mathbf{R}^2)$ has an unbounded component we proceed similarly. Since F^+ does not contain a halfspace there exists $\bar{\gamma}(0) = (\bar{x}(0), \bar{y}(0)) \in F^-$ with $\bar{x}(0) > x + Q$. Since $T_{-me_1} F^- \subseteq F^-$ we can construct a proper curve $\bar{\gamma} : [0, \infty) \to F^-$ such that $\bar{\gamma}(j) = \bar{\gamma}(0) - jme_1$ for all $j \in \mathbf{N}$. Then the same arguments as above show that $F^- \cap (\{x\} \times \mathbf{R}^2)$ has an unbounded component. □

Combining (4.6) and (4.9) we obtain:

LEMMA 4.10. *For every $\varepsilon > 0$ there exists a curve $\gamma : [0, \infty) \to F_\varepsilon$ with $\lim_{t \to \infty} |\gamma(t)| = \infty$.*

Proof. Choose $x \in \mathbf{R}$ according to (4.9), let Γ be an unbounded component of $F \cap (\{x\} \times \mathbf{R}^2)$ and let $\gamma : \mathbf{R} \to \Gamma$ be a diffeomorphism. Since F is properly embedded we have $\lim_{|t| \to \infty} |\gamma(t)| = \infty$. If $\varepsilon > 0$ is given choose $R > 0$ according to (4.6). There exists $t_0 \geq 0$ such that $|y(t)| > R$ if $|t| > t_0$ where $\gamma(t) = (x, y(t))$. Hence $\gamma(t) \in F_\varepsilon$ for $|t| \geq t_0$. □

In the outline of the proof the following fact about the involution $i : U \to U$ was important.

LEMMA 4.11. *For every $p \in U$ the points p and $i(p)$ lie in different components of U.*

Proof. Otherwise we can find a simple curve $\gamma : [0,1] \to U$ such that $i(\gamma(0)) = \gamma(1)$ and $\gamma(s) \neq i \circ \gamma(t)$ if $\{s,t\} \neq \{0,1\}$. Since i is an involution the curve

$\bar{\gamma} = \gamma * (i \circ \gamma) : [0,2] \to U$ is simple closed and thus bounds a disk D in F. On the other hand

$$h : (\mathbf{R}/2\mathbf{Z}) \times [0,2] \to F^+ \cup F$$
$$h(t,\tau) = \begin{cases} \bar{\gamma}(t) + \tau f(\bar{\gamma}(t)) N(\bar{\gamma}(t)) & \text{for } 0 \le \tau \le 1 \\ i \circ \bar{\gamma}(t) + (2-\tau) f(i \circ \bar{\gamma}(t)) N(i \circ \bar{\gamma}(t)) & \text{for } 1 \le \tau \le 2 \end{cases}$$

maps the cylinder $(\mathbf{R}/2\mathbf{Z}) \times [0,2]$ onto an embedded Möbius band $M \subset F^+ \cup F$ with $\partial M = F \cap M = \bar{\gamma}([0,2])$. Hence $M \cup D$ is a projective plane embedded into $\mathbf{R}^3$. Since the projective plane cannot be embedded into $\mathbf{R}^3$ a curve γ as above cannot exist.

Now we can complete the proof of Theorem (4.1). Since the principal curvatures of F are bounded there exists $r_0 \in (0,1)$ such that every metric ball $B_F(p,r_0)$ on F is strictly convex and can be represented as a graph over a topological disk in T_pF of a function with uniformly bounded gradient. In particular, there exist constants $A_1 = A_1(r_0) > 0$, $L_1 = L_1(r_0) < \infty$ such that

$$A(B_F(p,r_0)) \ge A_1, \quad L(\partial B_F(p,r_0)) \le L_1$$

for all $p \in F$. Now set $\delta = \min\{(5QL_1)^{-1}A_1, t_0\}$ where t_0 appears in the definition of $U = f^{-1}([0,t_0))$ and where $Q \ge 1$ is the quasiminimality constant, cf. (4.2). Finally choose $0 < \varepsilon < \min\{1, \frac{4\pi}{3}\delta^3\}$. Let $\gamma : [0,\infty) \to F_\varepsilon$ be a curve with $\lim_{t\to\infty} |\gamma(t)| = \infty$, cf. (4.10). Then there exists a sequence $t_i \to \infty$ such that $B_F(\gamma(t_i),r_0)$ and $B_F(\gamma(t_{i+1}),r_0)$ intersect in a single point while $B_F(\gamma(t_i),r_0) \cap B_F(\gamma(t_j),r_0) = \phi$ if $j > i+1$: we set $t_0 = 0$ and define inductively

$$t_i = \sup\{t | d_F(\gamma(t),\gamma(t_{i-1})) \le 2r_0\}.$$

We let $D_n \supset \bigcup_{i=0}^{n} B_F(\gamma(t_i),r_0)$ denote the topological disk which results from $\bigcup_{i=0}^{n} B_F$ $(\gamma(t_i),r_0)$ if one smoothes the cusps at the points in $B_F(\gamma(t_i),r_0) \cap B_F(\gamma(t_{i+1}),r_0)$, $0 \le i \le n-1$.

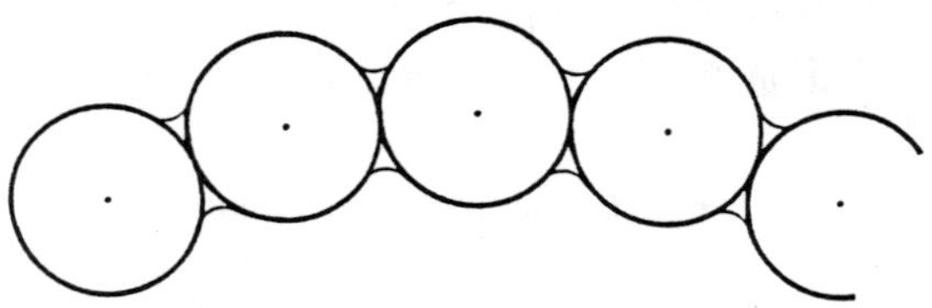

Figure 4

Then we have $A(D_n) \ge nA_1$ and $L(\partial D_n) \le nL_1$. According to (4.3) and since $r_0 < 1$ the function f satisfies $\frac{4\pi}{3} f^3(p) \le \varepsilon$, hence $f(p) \le \delta$, for all $p \in D_n$. In particular we have $D_n \subset U$. From (4.11) we know that D_n and iD_n are disjoint. The cylinder

$$Z_n : \partial D_n \times [0,1] \to \mathbf{R}^3, \quad Z_n(p,\tau) = \tau p + (1-\tau)i(p)$$

joining ∂D_n to $\partial(iD_n)$ has area $A(Z_n)$ satisfying

$$A(Z_n) \leq 4\delta L(\partial D_n)$$

since $f|\partial D_n \leq \delta$ and since i is locally Lipschitz with constant 2. Since the boundary of $\bigcup_{n\geq 0} D_n$ is connected we can find an arc Γ in $F \setminus \left(\bigcup_{n\geq 0} D_n\right)$ with endpoints $p_0 \in \partial D_0$ and $i(p_0) \in \partial(iD_0)$. Now we choose a fixed C^1-homotopy H from Γ to the line segment from $p_0 \in \partial D_0$ to $i(p_0)$ which lies on all the cylinders $Z_n, n \geq 0$. Since D_n and iD_n are disjoint disks on F the union of D_n, iD_n and a small tubular neighborhood of Γ is a disk in F. According to the "surgery" described in the outline of the proof this disk is fixed-boundary homotopic to a map into $\mathbf{R}^3$ with area bounded above by $A(Z_n) + 2A(H)$. Hence (4.3) implies

$$A(D_n) + A(iD_n) \leq Q(A(Z_n) + 2A(H)).$$

Hence we obtain the inequality

$$nA_1 < Q(4\delta n L_1 + 2A(H)).$$

For large n this contradicts our choice of δ. This contradiction concludes the proof of Theorem (4.1). □

REFERENCES

[1] D.V. Anosov, *On the behavior in the Euclidean or Lobachevsky plane of trajectories that cover trajectories of flows on closed surfaces I and II*, Izv. Akad. Nauk SSSR Ser. Mat. 51 (1987), pp. 16–43 and 52 (1988), pp. 451–478; English transl. in Math. USSR Izv. 30 (1988), pp. 15–38 and 32 (1989), pp. 449–474.

[2] D.V. Anosov, *Infinite curves on the torus and on closed surfaces of negative Euler characteristic*, Trudy Mat. Inst. Steklov 185 (1988), pp. 30–53; English transl. in Proc. Steklov Inst. Math. 185 (1990), pp. 33–58.

[3] V. Bangert, *A uniqueness theorem for $\mathbf{Z}^n$-periodic variational problems*, Comment. Math. Helv. 62 (1987), pp. 511–531.

[4] V. Bangert, *The existence of gaps in minimal foliations*, Aequationes Math. 34 (1987), pp. 153–166.

[5] V. Bangert, *On minimal laminations of the torus*, Ann. Inst. H. Poincaré – Analyse non linéaire 6 (1989), pp. 95–138.

[6] V. Bangert, *Laminations of 3-Tori by Least Area Surfaces*, In: Analysis, et cetera (P.H. Rabinowitz, E. Zehnder eds.), Academic Press 1990, pp. 85–114.

[7] M. Freedman, J. Hass, P. Scott, *Least area incompressible surfaces in 3-manifolds*, Invent. math. 71 (1983), pp. 609–642.

[8] C. Godbillon, Dynamical Systems on Surfaces, Springer 1983.

[9] M.W. Hirsch, Differential Topology, Springer 1976.

[10] J. Moser, *Minimal solutions of variational problems on a torus*, Ann. Inst. H. Poincaré-Analyse non linéaire 3 (1986), pp. 229–272.

[11] R. Schoen, *Estimates for stable minimal surfaces in three-dimensional manifolds*, Ann. of Math. Studies 103, Princeton Univ. Press, Princeton N.J. (1983), pp. 111–126.

[12] A. Weil, *On systems of curves on a ring-shaped surface*, J. Indian Math. Soc. 19 (1931), pp. 109–114.

THE ROTATION SET AS A DYNAMICAL INVARIANT

PHILIP BOYLAND*

Abstract. This paper discusses the question of how much information the rotation set of an annulus homeomorphism provides about the dynamics of the homeomorphism. In particular, for each number in the rotation set is there a "nice" minimal set with that rotation number? If the homeomorphism is area preserving monotone twist, the Aubry-Mather Theorem gives a very clear and complete answer to this question. Much of what is written here can be viewed as an attempt to generalize the results and techniques of area preserving monotone twist maps to the larger class of annulus homeomorphisms. There is also some discussion of related questions concerning what dynamics are possible when the rotation set is a single number. This paper surveys the current state of knowledge and raises a number of questions.

1. Introduction. The notion of rotation number goes back at least to Poincaré. One begins with a homeomorphism of the circle or annulus and iterates it to generate a dynamical system. The rotation number measures the asymptotic rate of rotation of orbits. For a homeomorphism of the circle, every point has the same rotation number and thus one speaks of the rotation number of the homeomorphism. For a homeomorphism of the annulus, different orbits can rotate at different asymptotic speeds and there is thus a rotation set. For a homeomorphism of the circle the rotation number tells one a great deal about the dynamics of the map. This paper concerns the question of how much information the rotation set gives about the dynamics of an annulus homeomorphism.

To view this question in a slightly larger mathematical context, note that in dynamics, conjugate maps are considered equivalent. The rotation set is unchanged by conjugacy and is thus an invariant of this equivalence relation. Given such an invariant, the natural mathematical question is the one raised above: How much information does the invariant give about the equivalence classes, i.e. about the dynamics of the maps. There is a related "realization question". If some real number r is contained in the rotation set, is it represented by a well understood, preferably indecomposable, invariant set which has that rotation number? To make an analogy, in topology one studies the equivalence classes of topological spaces under the equivalence relation of homeomorphism. The homology groups give an invariant of the equivalence classes. A standard question is how much does homology tell you about the underlying spaces? A typical realization question is whether subgroups are realized by embedded submanifolds.

Much of what is said below could be viewed as an attempt to generalize results that hold for area preserving, monotone twist maps to the wider class of annulus homeomorphisms. Two of the three standard avenues of approach to twist maps, variational and KAM, have nice generalizations to higher dimensions. The third, the topological approach, depends heavily on plane topology and seems to have few higher dimensional analogs. However, many of these topological techniques do have application to the class of annulus homeomorphisms. This has value and interest

*709 W. Nevada, Urbana, IL 61801

because many of the maps that arise in applications (for example, as cross sections or time-one maps of forced oscillators) satisfy neither the area preserving nor the monotone twist hypotheses.

The contents and purposes of this paper are at least three-fold. It is a survey of the current state of affairs, hopefully transporting useful information from the folklore into the literature. It contains a number of original remarks and it also raises a number of questions. The choice of results and questions represent, of course, only the personal idiosyncrasies of the author. Many similar issues and questions were raised by Hall in [Hl].

Acknowledgements. I would like to thank the following mathematicians for information and references; A. Fathi, G.R. Hall, M. Herman, A. Katok, R. MacKay, M. Turpin and S. Williams. I would also like to thank the IMA and the University of Minnesota for their support and hospitality.

2. Basic definitions and notation. Throughout this paper, the phrase "a homeomorphism f" refers to a homeomorphism $f : A \to A$ where $A = S^1 \times [0,1]$ is the annulus. We assume that f is isotopic to the identity, i.e. it preserves both orientation and boundary components. The phrase "area preserving" always means $f_* m = m$ where $dm = dr d\theta$, the coordinate measure on A. There are many interesting results and questions regarding the open annulus, preservation of an infinite area, area contracting maps, etc. However, for simplicity, we restrict our attention to the compact case with a smooth finite area.

The universal cover of A is $\widetilde{A} = \mathbf{R} \times [0,1], \pi_1 : \widetilde{A} \to \mathbf{R}$ is projection onto the first component and $\tilde{f} : \widetilde{A} \to \widetilde{A}$ denotes a lift of f. Given $\tilde{x} \in \widetilde{A}$, the lift of $x \in A$, define the *rotation interval* of x as

$$\rho(x) = \left[\liminf_{n\to\infty} \frac{\pi_1(\tilde{f}^n(\tilde{x})) - \pi_1(\tilde{x})}{n}, \limsup_{n\to\infty} \frac{\pi(\tilde{f}^n(\tilde{x})) - \pi_1(\tilde{x})}{n}\right]$$

Note that $\rho(x)$ is defined only up to choice of lift. If X is an invariant set, its rotation set is $\rho(X) = \{\rho(x) : x \in X\}$. The rotation set of f is $\rho(f) = \rho(A)$. Another popular definition of the rotation number is to define

$$\bar{\rho}(x) = \lim_{n\to\infty} \frac{\pi_1 \tilde{f}^n(\tilde{x}) - \pi_1(\tilde{x})}{n}$$

only when the limit exists. The rotation sets $\bar{\rho}(X) = \{\bar{\rho}(x) : x \in X\}$ and $\bar{\rho}(f) = \rho(A)$ are defined analogously. It is easy to show using results quoted below that $\rho(f) = \bar{\rho}(f)$.

A periodic point, x, is called a (p,q)*-periodic orbit* if x has least period q and $\rho(x) = p/q$. In several places below we shall have need to discuss the rate of convergence of the rotation number. If $\rho(x) = \alpha$, the *rotation number convergence function* is given by

$$r(x,n) = |\pi_1(\tilde{f}^n(x)) - \pi_1(\tilde{x}) - n\alpha|.$$

Clearly, $\rho(x) = \alpha$ implies $r(x,n) = o(n)$.

For $\alpha \in \mathbf{R}$, the map $\widetilde{R}_\alpha : \widetilde{A} \to \widetilde{A}$ is given by $\widetilde{R}_\alpha(x,y) = (x+\alpha, y)$. The projection is denoted $R_\alpha : A \to A$. The notation R_α will also be used for the analogous circle map.

Now a few notions from topological dynamics. A pair (W, g), with $g : W \to W$ a homeomorphism, is called *minimal* if for all $y \in W$, $\{g^n(y) : n \in \mathbf{Z}\}$ is dense in w. An invariant set $X \subset W$ is a *minimal set for g* if the pair $(X, g|_X)$ is minimal. We shall usually denote this pair by (X, g). Given another homeomorphism $h : Y \to Y$, one says that (Y, h) is a *factor of* (W, g) if there exists a continuous onto map $\varphi : W \to Y$ with $\varphi g = h\varphi$. More properly, this should be called a "topological factor" as there is an analogous measure theoretic notion.

3. Generalizations of the Aubry-Mather Theorem. Recall that a diffeomorphism $f : A \to A$ is called *monotone twist* if $\frac{\partial(\pi_1 \circ f)}{\partial y} > 0$. A f-invariant set $X \subset A$ is called *monotone* if $\tilde{f}$ restricted to the total lift $\widetilde{X}$ is order preserving, i.e. if $z_1, z_2 \in \widetilde{X}$ with $\pi_1(z_1) < \pi_1(z_2)$ then $\pi_1(\tilde{f}(z_1)) < \pi_1(\tilde{f}(z_2))$. For this to make sense we must require that π_1 restricted $\widetilde{X}$ be injective. The following theorem follows from the work of Aubry and of Mather.

THEOREM (Aubry-Mather). *Let f be an area preserving monotone twist map then*

(1) $\rho(f) = \left[\rho(S^1 \times \{0\}), \rho(S^1 \times \{1\})\right]$.
(2) *For all $p/q \in \rho(f)$ with p and q relatively prime, f has a monotone, (p,q)-periodic orbit.*
(3) *For all $\alpha \in \rho(f)$ with $\alpha \notin \mathbb{Q}$, f has a monotone, minimal set X with $\rho(X) = \alpha$ and (X, f) has (S^1, R_α) as a factor.*

This theorem is the inspiration for many of the questions we raise here. In a certain sense, it gives an ideal solution to the realization question raised in the introduction. For each number in the rotation set, the map has a minimal set with that rotation number. Further, since the minimal sets are monotone they are simplest in the sense that they are the same sort of minimal sets that occur for circle homeomorphisms. Thus the minimal sets given in part (3) are either invariant circles or Denjoy minimal sets.

Because the minimal sets are monotone and the map is monotone twist, one also gets information about how the minimal sets are embedded in the annulus. More precisely, these facts imply that the isotopy class of the map rel the invariant set is trivial in a certain sense. Thus in generalizing these results to arbitrary homeomorphism there are two cases, rational and irrational, and two questions in each case. For each rational p/q in the rotation set of f does f have a (p,q)-periodic orbit and is this periodic orbit embedded in the annulus like a monotone one? (The answer to both questions is yes.) For each irrational in the rotation set does f have a minimal set that is either an invariant circle or a Denjoy minimal set and if there is a minimal set with the given irrational rotation number is it nicely embedded in the annulus? (The answer to the first question is no and for the second, existence and embedding are not yet understood.) The discussion of these questions is the purpose of this section.

There are basic results regarding these questions due to Handel ([Hn2]). First, $\rho(f)$ is always a closed set, and second, for all but at most a finite number of $r \in \rho(f)$, f has a minimal set X with $\rho(X) = r$. It is easy to show that any compact subset of $\mathbf{R}$ is $\rho(f)$ for some homeomorphism f.

Question 3.1: For generic f, does $\rho(f)$ have finitely many components?

3A: Rationals in the rotation set. In generalizing part (2) of the Aubry-Mather Theorem the first result needed is that $p/q \in \rho(f)$ with p and q relatively prime implies that f has a (p,q)-periodic orbit. This was proved by Franks ([F]) and Handel ([Hn3]). The next step is to analyze how the periodic orbit is embedded in the annulus. For a general homeomorphism, the notion of monotone is much too restrictive. The appropriate generalization of monotone uses an idea of Bowen ([B]) and characterizes the periodic orbit by the action of the map on its complement.

DEFINITION. A (p,q)-periodic orbit, $o(x,f)$, is said to be *topologically monotone* if there exists a homeomorphism $h : A \to A$ and a periodic orbit $o(y, R_{p/q})$ with $h(o(x,f)) = o(y, R_{p/q})$ and $h \circ f \circ h^{-1}$ isotopic to $R_{p/q}$ rel $o(y, R_{p/q})$.

One has in analogy with part (2) of Aubry-Mather:

THEOREM ([Bd]). *For all $p/q \in \rho(f)$ with p and q relatively prime, f has a topologically monotone (p/q)-periodic orbit.*

Under the area preserving hypothesis this theorem was first proved by LeCalvez ([LC]). His paper contains powerful ideas which allow one to transport many ideas from the study of monotone twist maps to the study of general annulus homeomorphisms.

3B: Irrationals in the rotation set. Using the Aubry-Mather Theorem as a guide, one would hope for the existence of a minimal set X which has (S^1, R_α) as a factor for each irrational $\alpha \in \rho(f)$. This is not always the case, as was shown by an example of Handel ([Hn1]). This example is a C^∞-area preserving diffeomorphism, h. There is an irrational $\alpha \in \rho(h)$ and a minimal set, X, so that every point with rotation number α is contained in X. The set X is topologically a pseudocircle and (X,h) *does not have* (S^1, R_α) as a factor. This raises the question of what factors are needed for a general homeomorphism. This is formalized as follows:

DEFINITION. A collection of minimal sets $\mathcal{M} = \{(X_\lambda, g_\lambda) : \lambda \text{ is in an index set } \Lambda\}$ is called *collection of minimal α-models* if for any homeomorphism f with $\alpha \in \rho(f)$ there exists a minimal set $X \subset A$ and a unique $\lambda \in \Lambda$ such that (X,f) has (X_λ, g_λ) as a factor.

Question 3.2: Describe a collection of minimal models.

It should be remarked that at the current state of knowledge it is conceivable that there does not exist a collection of minimal models, i.e. there could be a homeomorphism f with $\alpha \in \rho(f)$ and f has no minimal sets all of whose points have rotation number α. This seems somewhat unlikely (but see the remark below on the rotation sets of minimal sets). Even more unlikely, but also conceivable, is the possibility that the structure of a collection of minimal α-models could depend on the arithmetic properties of α.

Another interesting possibility is raised by the following property of Handel's example. Points on both sides of the pseudo circle have smaller rotation numbers than the pseudo circle. This means that the example does not exclude a positive answer to the following question. A positive answer would provide a generalization of the Aubry-Mather Theorem that holds for boundary twist maps. These are the type of homeomorphisms considered in the classical Poincaré-Birkhoff theorem.

Question 3.3: Let f be an area preserving homeomorphism. For each irrational, α, which lies between the rotation numbers of f restricted to the boundary circles, does f have a minimal set X so that (X, f) has (S^1, R_α) as a factor?

Thus far we have concentrated on finding a minimal set to "represent" each $\alpha \in \rho(f)$. A weaker representation would be a point x with $\rho(x) = \alpha$ and this rotation number converges as quickly as possible. More precisely,

Question 3.4: If $\alpha \in \rho(f)$ does there always exist a point $\tilde{x} \in \widetilde{A}$ with $\rho(\tilde{x}) = \alpha$ and whose rotation number convergence function (defined in Section 2) satisfies $r(\tilde{x}, n) = O(1)$?

For a monotone twist f, any point x contained in the monotone minimal set given by the Aubry-Mather Theorem does satisfy $r(x, n) = O(1)$.

3C: Rotation sets of invariant sets. This section is concerned with generalizations of part (i) of the Aubry-Mather theorem. We interpret this in a very broad sense as dealing with the rotation set of various invariant sets. The following theorem is of central importance. It follows from combining a result of Franks ([F]) with the fact that $\rho(f)$ is closed ([Hn2]).

THEOREM. *If X is a chain transitive, compact invariant set then convexhull $(\rho(X)) \subset \rho(f)$.*

Thus, for example, if f is area preserving then $\rho(f)$ is a closed interval. One also has that $p/q \in$ *convexhull* $(\rho(X))$ implies that f has a (p, q)-periodic orbit. Note, however, that the theorem does not say that this periodic orbit is contained in X. We shall see below an example of a minimal set X with $\rho(X) = 1/2$ but X is not a periodic orbit. One does expect that "dynamically indecomposable" sets should have indecomposable rotation sets. This is expressed in the next question. For these questions the rotation set defined using the limit, $\bar{\rho}$, is more appropriate.

Question 3.5: If X is a minimal set or a chain transitive set or a connected component of the chain recurrent set, is $\bar{\rho}(X)$ connected? If X is compact, is $\bar{\rho}(X)$ compact? In these cases, is $\bar{\rho}(X) = \rho(X)$?

There are some interesting results related to this question in a paper by Barge and Swanson ([BS]). Their paper also develops the connection of rotation number to some standard objects and results in ergodic theory. Given the lift of a homeomorphism $\tilde{f}$, if we define $\varphi : \widetilde{A} \to \mathbf{R}$ by $\varphi(z) = \pi_1(\tilde{f}(z)) - \pi_1(z)$ then

$\tilde{\rho}(\tilde{x}, \tilde{f}) = \lim_{m\to\infty} \sum_{n=0}^{m} \frac{\varphi(\tilde{f}^n(\tilde{x}))}{m+1}$ if the limit exists. Thus the rotation number is just a Birkhoff average. From this point of view, Question 3.5 concerns the set of Birkhoff averages of a particular function φ. This involves understanding not only points that are generic for some ergodic measure μ supported on X (i.e. points with $\rho(x, f) = \int \varphi d\mu$) but also the more delicate issue of nongeneric points whose ergodic average of φ happens to converge.

These considerations will be useful in discussing the examples below. To construct these examples, we shall need a well known technique for connecting symbolic dynamics to rotation number (cf [Bh]). The basic idea when used in the annulus goes back at least to Levi ([L]).

The main idea is easily described in terms of the circle map ψ whose graph (mod 1) is given below in figure 1.

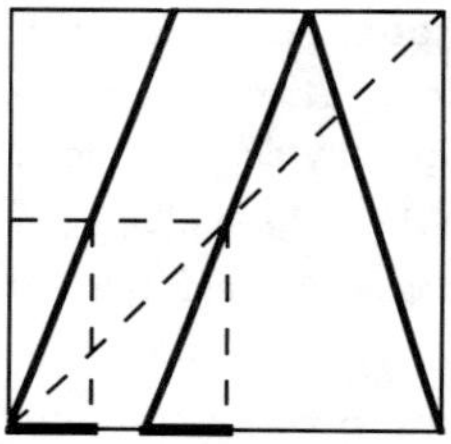

Figure 1.

The nonwandering set of ψ restricted to the union of the intervals A and B (as shown) is topologically conjugate to the full shift $\{0,1\}^{\mathbf{Z}}$. Further, if $s = \ldots s_{-1}s_0s_1\ldots$ is the sequence corresponding to a point $x \in S^1$ then

$$\bar{\rho}(x, \psi) = \lim_{m\to\infty} \frac{\sum_{n=0}^{m} s_n}{m+1}.$$

In other words, the rotation number of x is just the asymptotic number of ones in the symbol s. Call the right hand side of this equation $A(s)$. In the language introduced above, $A(s)$ is the Birkhoff average of the indicator function of $\{s \in \{0,1\}^{\mathbf{Z}} : s_0 = 1\}$ under the shift automorphism.

One can easily accomplish an analogous construction in the annulus by "fattening things up". This produces an Axiom A diffeomorphism of the annulus, Φ, with a compact invariant set, Λ, with the property that Φ restricted to Λ is topologically conjugate to the full shift. Further, if s is the sequence that corresponds to $x \in \Lambda$ then $\rho(x) = A(s)$.

This correspondence allows us to transfer well known examples from the two shift into the annulus. For example, the Morse minimal set, M, ([GH] pg. 105–111) is easily seen to have the property that $A(s) = 1/2$ for all $s \in M$. (This also follows from the fact that M is strictly ergodic as was proved by Kakutani.) Using the topological conjugacy above, this gives a minimal set X_M for Φ with $\rho(X_M) = 1/2$ but X_M is not a periodic orbit.

One can also use this construction to produce minimal sets for annulus homeomorphisms with nontrivial rotation sets. The basic observation is that minimal sets are not always uniquely ergodic, i.e. they may support many ergodic measures. Examples in the two shift began with Oxtoby ([O], see also [W]). If a minimal set $Z \subset \{0,1\}^{\mathbf{Z}}$ supports two ergodic measures that assign different masses to the set $\{s \in \{0,1\}^{\mathbf{Z}} : s_0 = 1\}$ then using the Birkhoff ergodic theorem, $\{A(s) : s \in z\}$ contains at least two points. Using the construction above, one then has a minimal set for a homeomorphism of the annulus which contains points with different rotation numbers. Susan Williams has constructed examples of minimal sets which have a nontrivial closed interval of rotation numbers (personal communication).

One can also use this construction to produce points whose rotation numbers converge as slowly as desired. More precisely, let $\varphi : \mathbf{N} \to \mathbf{R}^+$ satisfy $\frac{\varphi(n)}{n} \to 0$. For simplicity, assume that φ is monotone increasing, $\varphi(n) \to \infty$ and there exists an N so that $n > N$ implies $\varphi(n+1) - \varphi(n) < 1$.

One can then find a point x whose rotation number convergence function under Φ satisfies $\frac{r(x,n)}{\varphi(n)} \to 1$. To see this, for $s \in \{0,1\}^{\mathbf{Z}}$ with $A(s) = \alpha$ define $r(s,n) = \left|\sum_{i=1}^{n} s_i - n\alpha\right|$. It clearly suffices to find an s with $\frac{r(s,n)}{\varphi(n)} \to 1$.

To construct such an s we begin with the simplest so called Sturmian sequence (see [GH], pg. 111). Pick an irrational α with $0 < \alpha < 1/4$ and define s by $s_n = 1$ if the fractional part of $n\alpha$ is between zero and α and $s_n = 0$, otherwise. The sequence s has the properties that $A(s) = \alpha$, $r(s,n) \leq 1$ and s does not contain a pair of consecutive ones. It is worthwhile noting (although we shall not use it) that the closure of the orbit of s is conjugate to a Denjoy minimal set.

Given the function $\varphi(n)$, construct a new sequence t as follows. Let $\varepsilon_i = 0$ or 1 and be such that $\sum_{i=1}^{n} \varepsilon_i = \lfloor \varphi(n) \rfloor$ where $\lfloor \ \rfloor$ means integer part. If $\varepsilon_i = 0$, let $t_i = s_i$. If $\varepsilon_i = 1$ and $s_i = 0$ let $t_i = 1$ and if $s_i = 1$ let $t_i = 1$ and $t_{i+1} = 1$. Thus, $r(t,n) = |r(s,n) + \lfloor \varphi(n) \rfloor + \delta_n|$ with $\delta_n = 0$ or 1 and so $\frac{r(t,n)}{\varphi(n)} \to 1$ as required.

It is a somewhat curious phenomenon that the existence of a certain type of minimal set can force the existence of periodic orbits in its complement. For example, if f possesses the Morse minimal set M with $\rho(M) = 1/2$ then f must have a (1,2)-periodic orbit. An even more striking example is the fact that any periodic orbit which has a nontrivial action on its complement implies the existence of infinitely many periodic orbits. This is expressed in

THEOREM ([Bd]). *If p and q are relatively prime and f has a (p,q)-periodic orbit that is not topologically monotone then the Farey interval of p/q is contained in $\rho(f)$. (The Farey interval of p/q is the closed interval whose end points are the nearest rationals with denominators less than q.)*

Two final remarks on the theme of this section: If $\text{Fix}(f^n)$ is finite for all n and f has a point x with $\rho(x)$ a nontrivial interval then f has positive topological entropy ([Hn3]). The second remark concerns Caratheodory's theory of prime ends. It seems likely that this theory could shed considerable light on many questions related to the theme of this section. Indeed, recent work Alligood, Sauer and Yorke, Barge, Gillette and Swanson, and LeCalvez indicate this is so. Regrettably, there is insufficient time to discuss this interesting work.

4. The rotation set is a single point. In understanding the implications of the rotation set for the dynamics, it's clear that one must confront the case where $\rho(f)$ is a single point. Before looking at the two cases, rational and irrational, we raise a general question regarding the rates of convergence of rotation numbers. Recall that the rotation number convergence function $r(\tilde{x}, n)$ was defined in section 2.

Question 4.1: If the rotation set of f is a single point, how much can the asymptotic behavior of $r(\tilde{x}, n)$ vary as $\tilde{x}$ varies?

4A: The rotation set is a single rational number. The simplest examples of such maps are obtained by partitioning the annulus into q sectors, each of the form $S_i = \left\{(r, \theta) : \frac{2\pi i}{q} \leq \theta \leq \frac{2\pi(i+1)}{q}\right\}$. In sector S_0, one puts a homeomorphism of the disk that fixes the boundary and in the other sectors, one has the identity. This map is then composed with rigid rotation by p/q. This construction can be used to produce homeomorphisms with $\rho(f) = p/q$ which have a dense orbit or even are Bernoulli (see [K]).

If $\rho(f) = p/q$ we know there must be at least one (p, q)-periodic orbit. Can one produce examples with little periodic behavior?

Question 4.2: If $\rho(f) = p/q$ and f is area preserving, must f have infinitely many periodic orbits?

This is related to a question of Herman which asks whether an area preserving homeomorphism of the three sphere that has 3 fixed points must in fact have infinitely many periodic points. Another related question is: Does an area preserving homeomorphism of the disk that is the identity on the boundary always have infinitely many interior periodic points?

A particularly restrictive case in which $\rho(f) = p/q$ is when $f^q = id$. In this case a result of Brouwer ([Bw]), Kerekjarto ([Kj]) and Eilenburg ([E]) states that f is topologically conjugate to $R_{p/q}$.

4B: The rotation set is a single irrational number. The simplest examples here are homeomorphisms all of whose recurrence lies on homotopically nontrivial invariant circles with the irrational rotation number. To restrict the field of examples somewhat we focus on the area preserving case.

THEOREM (Franks [F]). *Let f be chain transitive (for example, f is area preserving) then f has no periodic orbits if and only if $\rho(f) = \alpha \notin \mathbb{Q}$.*

The simplest example of such a map is R_α. The question of whether this is the only example goes back to Birkhoff ([Bk]).

> Problems 14 and 15 were closely related. The first of these asserts that a 1-1 direct analytic area preserving deformation of the surface of a sphere into itself which has two fixed points, and is such that iterates of the transformation have no other fixed points, is a pure rotation from the topological point of view. Considerable evidence

was adduced for this conjecture. The second problem embodied an analogous conjecture concerning a plane circular ring.

DEFINITION. An area preserving homeomorphism $f : A \to A$ is called a *fake rotation* if $\rho(f) = \alpha \notin \mathbf{Q}$ and f *is not* topologically conjugate to R_α.

Question 4.3: For which values of k and α do there exist C^k-fake rotations with rotation number α?

In this terminology we have:

Birkhoff Conjecture: There do not exist C^ω-fake rotations.

To the best of my knowledge, the Birkhoff conjecture is still open. However, a fair amount of information is available about Question 4.3. Using methods of Anosov and Katok ([A-K]) one can construct C^∞ fake rotations that have dense orbits (see appendix). Katok pointed out that these methods also produce examples that are not even measure isomorphic to R_α. There are related examples and results by Snirelman ([Sn]), Besicovitch ([Bs1]), Sidorov ([Sd1]), Oxtoby([O]) Herman ([Hm1]) and Fokkink & Oversteegen ([FO]).

If we only require that that f preserves a measure equivalent to the coordinate measure then one can produce C^ω-maps with a dense orbit and no periodic orbits. This was essentially done by Sidorov ([Sd1]). Here's an example due to Herman that is a variant on the skew product construction.

The first step is to use standard techniques involving lacunary Fourier series to produce a Liouville $\alpha \notin \mathbf{Q}$ such that $\int f\, d\theta = 0$ and the equation

$$(*) \quad g(\theta + \alpha) - g(\theta) = f(\theta)$$

has a solution $g \in L^2(S^1)$ but no continuous solutions. (This is done explicitly in [Mn].)

Now if we define $\Phi : A \to A$ via $\Phi(\theta, r) = (\theta + \alpha, \phi_{f(\theta)} r)$ where ϕ_t is the solution flow to $\dot{r} = r(1-r)$ then Φ is a C^ω-diffeomorphism. To see that Φ has a dense orbit, first define $h : S^1 \times \mathbf{R} \to S^1 \times (0,1)$ via $h(x,y) = (x, \phi_y(1/2))$ and $T : S^1 \times \mathbf{R} \to S^1 \times \mathbf{R}$ via $T(x,y) = (x + \alpha, y + f(x))$. A simple computation shows that $h^{-1}\Phi h = T$. Because equation $(*)$ has no continuous solution g for our choice of f, a theorem of Gottschalk and Hedlund ([GH], page 136) implies that T and thus Φ is transitive (i.e. it has a dense orbit).

On the other hand, if we let g be an L^2 solution to $(*)$ and define $\beta : A \to A$ by $\beta(\theta, r) = (\theta, \phi_{g(\theta)}(r))$ then $\beta^{-1}\Phi\beta = R_\alpha$ so Φ preserves $\beta_*(m)$.

Anosov ([A]) gives some interesting history of the skew product construction. The basic idea goes back to Poincaré and Hilbert mentioned equation $(*)$ in the description of his fifth problem.

All these examples appear to have rotation numbers that are well approximated by rationals. Little is known about the relationship of smoothness and rotation number for fake rotations (if, in fact, there is one). Much is known, however, about the smoothness of solutions to equation $(*)$ for various α and thus about the skew product construction (see [Hm2] and the references therein). In particular, the

C^ω- construction done above can only be done for (super) Liouville α. However, using a result attributed to Y. Meyer in [Hm2] (pg 187) one can construct C^1-diffeomorphisms with any irrational rotation number that have dense orbits and no periodic points (but perhaps not preserving the coordinate measure).

All the constructions considered above (at least when they are done in the simplest way) produce homeomorphisms that are *recurrent* in the following sense: For all $\varepsilon > 0$, there exists an n so that $d(f^n, Id) < \varepsilon$ in the C^0-topology. If the collection of iterates of f is an equicontinuous family then f is called *almost periodic*. Clearly almost periodic implies recurrent. It is obviously a much strong condition, in fact, Hemmingson ([Hg]) has shown that there are no almost periodic fake rotations.

We end this section with a question that seems to have first appeared in Besicovitch ([Bs2]). Recall that a homeomorphism $h : X \to X$ is called minimal (respectively, semiminimal) if every orbit (forward orbit) is dense. If X is compact, these two notions are equivalent. However, Gottschalk showed that in a locally compact metric space every semiminimal set is compact ([G], Theorem B). Thus, in particular, if X is not compact, it has no semiminimal self-homeomorphism. Further, if a noncompact, locally compact metric space X has a minimal homeomorphism, there must be a point x whose orbit converges to infinity in forward time. This follows from the Gottschalk result by observing that there must be a point x whose forward orbit is not dense. The omega limit set of x must be "infinity" in the one point compactification as the homeomorphism is minimal and an omega limit set is a closed invariant set.

This is a well known remark whose origin I have been unable to trace. However, there is a closely related result due to Homma and Kinoshita ([HK]): If X is a locally compact, noncompact, separable metric space and $f : X \to X$ is continuous then $Q(f) = \{x \in X :$ the forward orbit of X is not dense $\}$ is a dense set. If $X - Q(f)$ is nonempty (i.e. f has a dense forward orbit) then it is easy to show that $X - Q(f)$ is also dense in X.

In [Bs2], the question is raised whether the once punctured plane possesses a minimal homeomorphism. This seems to be still open. A closely related question can be raised for the annulus. Say a homeomorphism $f : A \to A$ is minimal if f restricted to Int (A) is minimal. By the observations above there are no semiminimal homeomorphisms of the annulus.

Question 4.4: (Herman) Do there exist minimal homeomorphisms of the annulus?

Note that a minimal homeomorphism, f, is chain transitive and has no periodic orbits and thus by the result of Franks stated at the beginning of the section, $\rho(f) = \alpha \notin \mathbb{Q}$. Note also that by the consequence of Gottschalk's theorem stated above, a homeomorphism of A cannot be both minimal and recurrent.

5. Invariant circles for monotone twist maps. There are many interesting open questions regarding twist maps. We only raise two that fit in well with our general theme. *The convention of this section is that $g : A \to A$ is always an area preserving, C^∞-monotone twist map.* Define $\rho c(g) = \{\rho(\Gamma) :$

Γ is a homotopically nontrivial invariant circle for g }. It is well known that the set $\rho c(g)$ is compact.

Question 5.1: Find conditions on a compact set $B \subset \mathbf{R}$ which are necessary and sufficient for the existence of a g with $B = \rho c(g)$.

This is a difficult problem. It includes such deep issues as the rotation numbers of boundary circles and isolated circles. This is a good deal of information in the Physics literature about aspects of this question. For a start the reader is referred to [GMS] and [M]. The recent remarkable work of Herman and Yoccuz on generic invariant circles in the closure of the KAM circles is also relevant to this question. For example, the generic such circle has Liouville rotation number.

We close with an even more difficult but equally fascinating question.

Question 5.2: If g_μ is a generic one parameter family with g_0 integrable, describe the evolution of $\rho c(g_\mu)$ as μ increases from zero.

Appendix

Some years ago, M. Herman pointed out that the methods of Anosov and Katok ([AK]) could be used to provide a positive answer to Birkhoff's question for C^∞-diffeomorphism as phrased in Question 4.3 above. In particular, a minor alteration of the proofs in [FH] would suffice. This is what is presented here.

The purpose of this appendix is to present the powerful techniques of [AK] and their elegant application in [FH] in what is perhaps the simplest, nontrivial case. For a history of this circle of ideas, the reader is referred to the introduction of [AK]. The arguments here follow [FH] very closely.

We maintain the notation defined above, in particular, "area" always means the coordinate measure. Let $\mathcal{P}$ be the space of all area preserving, C^∞-diffeomorphisms of A that are isotopic to the identity. The space $\mathcal{P}$ is given the C^∞-topology. For $\alpha \in S^1$, recall that $R_\alpha : A \to A$ is defined by $R_\alpha(r,\theta) = (r, \theta + \alpha)$. Let $\Lambda = \{g^{-1}R_\alpha g : g \in \mathcal{P} \text{ and } \alpha \in S^1\}$ and $\overline{\Lambda}$ be its closure. Note that $\overline{\Lambda}$ is a Baire space.

A diffeomorphism $f \in P$ is said to be *almost minimal* if almost every point in A has a dense orbit. This happens if and only if every compact invariant set has zero or full measure. Note that an example of Sidorov ([Sd2]) shows that almost minimal is stronger than topologically transitive. Finally, say that f is *smoothly recurrent* if there is a sequence $n_i \to \infty$ with $f^{n_i} \to Id$ in the C^∞-topology.

PROPOSITION. *The subset of $\overline{\Lambda}$ consisting of almost minimal, smoothly recurrent diffeomorphisms with no periodic orbits is dense G_δ in $\overline{\Lambda}$.*

Remark. Any such diffeomorphism provides an answer to Question 4.3 for C^∞-diffeomorphisms.

Proof. If $N_n = \{f \in \overline{\Lambda} : \text{Fix}\,(f^n) = \emptyset\}$ then N_n is open and dense in $\overline{\Lambda}$ since $\{gR_\eta g^{-1} : g \in \mathcal{P} \text{ and } \eta \notin \mathbf{Q}\} \subset N_n$. Since $\bigcap_{n\in\mathbf{N}} N_n$ is precisely the maps in $\overline{\Lambda}$ with no periodic points, this set of maps is dense G_δ in $\overline{\Lambda}$. Similarly, if $R_n = \{f \in \overline{\Lambda} : \exists k \in \mathbf{N} \text{ such that } d(f^k, id) < \frac{1}{n} \text{ in the } C^\infty\text{-topology}\}$ then R_n is open and dense. The set of smoothly recurrent maps in $\overline{\Lambda}$ is precisely $\bigcap_{n\in\mathbf{N}} R_n$ and so this set is also dense G_δ.

It remains to show that almost minimal is also dense G_δ in $\overline{\Lambda}$. Let $U \subset A$ be an open set and $n \in \mathbf{N}$. Define $M(U,n) = \{f \in \overline{\Lambda} : \exists k \in \mathbf{N}$ with $m(\bigcup_{j=0}^{k} f^{-j}(U)) > 1 - 2/n\}$. If $\{U_i\}$ is a countable base for the topology of A, it is easy to show that $\bigcap_{i,n} M(U_i, n)$ is the set of almost minimal diffeomorphisms in $\overline{\Lambda}$. Each $M(U,n)$ is clearly open so the proof is complete after showing that each $M(U,n)$ is dense in $\overline{\Lambda}$. For this we need the following lemma:

LEMMA. *Given $n \in \mathbf{N}$ and a connected, nonempty open set $U \subset A$ then there exists an $H \in \mathcal{P}$ such that $H(U)$ connects $[-1, -1+1/n] \times S^1$ and $[1-1/n, 1] \times S^1$. Further, given $p/q \in \mathbf{Q} \cap [0,1]$ we may chose H so that $HR_{p/q}H^{-1} = R_{p/q}$.*

Proof. The first part is an easy exercise. For the second, treat A as a q-fold cover of $A/R_{p/q}$. The required H is the lift to A of the diffeomorphism of $A/R_{p/q}$ obtained by applying the first part to $\pi(U)$ where $\pi : A \to A/R_{p/q}$ is the projection. □

Now suppose n and U are given. Construct H as in the lemma. It's clear that for any $\eta \notin \mathbf{Q}$ there exists a k so that

$$m\left(\bigcup_{j=0}^{k} R_\eta^{-j}(H(U))\right) > 1 - 2/n$$

or since H^{-1} preserves the measure

$$m\left(\bigcup_{j=0}^{k} H^{-1}R_\eta^{-j}H(u)\right) > 1 - 2/n.$$

Thus $H^{-1}R_\eta H \in M(U,n)$. This remark will be used shortly.

We are now in a position to show that each $M(U,n)$ is dense. Since by definition, $\overline{M(U,n)} \subset \overline{\Lambda}$ it suffices to show that $\Lambda \subset \overline{M(U,n)}$. Now given $g^{-1}R_\alpha g \in \Lambda$, using the fact that g^{-1} preserves the measure it's clear that $g^{-1}R_\alpha g \in \overline{M(U,n)}$ if and only if $R_\alpha \in \{f \in \overline{\Lambda} : \exists k \in \mathbf{N}$ with $m\left(\bigcup_{j=0}^{k} f^{-j}(g(U)\right) > 1 - 2/n\}$. Thus it suffices to show that for all $\alpha \in S^1$, open, connected U and positive integers n, $R_\alpha \in \overline{M(U,n)}$.

Now fix p/q and chose $\eta_n \notin \mathbf{Q}$ with $\eta_n \to p/q$. Using the lemma and the remark after it, find an $H \in \mathcal{P}$ with $H^{-1}R_{p/q}H = R_{p/q}$ and $H^{-1}R_\eta H \in M(U,n)$ for all $\eta \notin \mathbf{Q}$. We thus have $H^{-1}R_{\eta_n}H \to H^{-1}R_{p/q}H = R_{p/q}$ and so $R_{p/q} \in \overline{M(U,n)}$. If $\alpha \notin \mathbf{Q}$, pick $p_n/q_n \to \alpha$ and so $R_{p_n/q_n} \to R_\alpha$ which is thus in $\overline{M(U,n)}$.

Remark. Recall that $f : A \to A$ is called minimal, if for all x in the interior of A, the set $\{f^n(x) : n \in \mathbf{Z}\}$ is dense. As noted at the end of section 4, a recurrent homeomorphism cannot be minimal. Thus, the generic map in $\overline{\Lambda}$ is not minimal.

REFERENCES

[A] Anosov, D., *On an additive functional homology equation connected with an ergodic rotation on a circle*, Math. USSR Izv., 7 (1973), pp. 1257–1271.

[AK] Anosov, D. and Katok, A., *New examples in smooth ergodic theory*, Trans. Moscow Math. Soc., 27 (1972), pp. 117–134.

[BS] Barge, M. and Swanson, R., *Rotation shadowing properties of circle and annulus maps*, Ergod. Th. & Dynam. Sys., 8 (1988), pp. 509–521.

[Bs1] Besicovitch, A., *A problem on topological transformations of the plane*, Fund. Math., 28 (1937), pp. 61–65.

[Bs2] Besicovitch, A., *A problem on topological transformations of the plane, II*, Proc. Cambridge Philos. Soc., 47 (1951), pp. 38–45.

[Bk] Birkhoff, G.D., *Some unsolved problems in theoretical dynamics*, reprinted in Collected Paper, vol II, pp. 710–712.

[Bh] Bothelo, F., *Rotation sets for annulus homeomorphisms*, Pac. J. of Math., 133 (1988), pp. 251–266.

[Bd] Boyland, P., *Rotation sets and topologically monotone orbits for annulus homeomorphisms*, preprint.

[B] Bowen, R., *Entropy and the fundamental group*, in Springer LNIM, 668.

[Bw] Brouwer, L., *Über die periodischen transformationen der Kugel*, Math. Ann., 80 (1919), pp. 39–41.

[E] Eilenburg, S., *Sur les transformations periodiques de la surface de sphere*, Fund. Math., 22 (1934), pp. 28–44.

[FH] Fathi, A. and Herman, M., *Existence de diffeomorphismes minimaux*, Astérisque, 49 (1977), pp. 37–59.

[FO] Fokkink, R. and Oversteegen, L., *An example related to the Birkhoff conjecture*, (preprint).

[F] Franks, J., *Recurrence and fixed points of surface homeomorphisms*, Ergod. Th. Dynam. Sys., 8* (1988), pp. 99–107.

[G] Gottschalk, W., *Orbit-closure decompositions and almost periodic properties*, Bull. AMS, 50 (915–919).

[GH] Gottschalk, W. and Hedlund, *Topological Dynamics*, AMS Colloquium Pub., vol. 36 (1955).

[GMS] Greene, J., MacKay, R., and Stark, J., *Boundary circles for area preserving maps*, Physica 21 D (1986), pp. 267–295.

[Hl] Hall, G.R., *Some problems on dynamics of annulus maps*, Contemp. Math., 81 (1989), pp. 135–152.

[Hn1] Handel, M., *A pathological C^∞-diffeomorphism of the plane*, Proc. AMS, 86 (1982), pp. 163–168.

[Hn2] Handel, M., *The rotation set of a homeomorphism of the annulus is closed*, Commun. Math. Phys., 127 (1990), pp. 339–349.

[Hn3] Handel, M., *Zero entropy surface homeomorphisms*, preprint.

[Hg] Hemmingson, E., *Plane continua admitting non-periodic autohomeomorphisms with equicontinuous iterates*, Math. Scand., 2 (1954), pp. 119–141.

[Hm1] Herman, M., *Construction of some curious diffeomorphisms of the Riemann sphere*, J. London Math. Soc, 34 (1986), pp. 375–384.

[Hm2] Herman, M., *Sur la conjugaison différentiable des différentiable des difféomorphismes du cercle à des rotations*, Publ. IHES, 49 (1979).

[HK] Homma, T. and Kinoshita, S., *On the regularity of homeomorphisms of* $\mathbf{E}^n$, J. Math. Soc. Jap., 5 (1953), pp. 365–371.

[K] Katok, A., *Bernoulli diffeomorphisms on surfaces*, Annals of Math., 110 (1979), pp. 529–547.

[Kj] Kerekjarto, B., *Sur la structure des topologigues des surfaces en elles-memes*, L'Enseignement Math., 35, pp. 297–316.

[LC] LeCalvez, P., *Existence d'orbites de Birkhoff generalisées pour les difféomorphisme conservatifs de l'anneau*, preprint.

[L] Levi, M., *Qualitative analysis of the periodically forced relaxation oscillation*, Mem. AMS, 244.

[M] MacKay, R., *Transition to chaos in area preserving maps*, in Springer Lecture Notes in Physics, 247.

[Mn] Mane, R., Ergodic Theory and Differentiable Dynamics, Springer Verlag (1987).

[O1] OXTOBY, J., *Ergodic sets*, Bull. A.M.S., 58 (1952), pp. 116–136.

[O2] OXTOBY, J., *Note on transitive transformation*, Proc. Nat. Acad. Sci. USA (1937), pp. 443–446.

[OU] OXTOBY, J. AND ULAM, S., *Measure preserving homeomorphisms and metrical transitivity*, Annals of Math., 42 (1941), pp. 874–920.

[Sd1] SIDOROV, E., *Smooth topologically transitive dynamical systems*, Math. Notes, 4 (1968), pp. 939–943.

[Sd2] SIDOROV, E., *Topologically transitive cylindrical cascades*, Math. Notes, 14 (1973), pp. 810–816.

[Sn] SNIRELMAN, L., *Example of a transformation of the plane*, Izv. Donsk. Politehn. Inst. Novocerkasske 14, Nauen. Otdel. Fiz-Mat. Cast, (Russian) (1930), pp. 64–74.

[W] WILLIAMS, S., *Toeplitz minimal flows that are not uniquely ergodic*, Z. Wahrscheinlichkeittheorie verw. Gebiete, 67 (1984), pp. 95–107.

BIRKHOFF PERIODIC ORBITS FOR SMALL PERTURBATIONS OF COMPLETELY INTEGRABLE HAMILTONIAN SYSTEMS WITH NONDEGENERATE HESSIAN

WEIFENG CHEN*

1. Introduction. Originally, the motivation of this work was intended to generalize the Aubry-Mather theory to a high dimension system, and it is part of our program, see [BK], [M], [K]. Recently M.Herman's work [H1] indicates that it is impossible to get a reasonable analog of the Aubry-Mather theory in a high dimension system if no further structure is imposed. But our result is still very interesting for the following reasons:

(1) It is a high dimension analog of Poincaré's work on his Poincaré's Last Geometric Theorem. Remember that Poincaré proved his celebrated theorem only in the near integrable case.

(2) We greatly improved the results obtained in [BK] . We abandoned the convexity condition completely and only need the nondegeneracy condition for the generating function, which is the standard condition for the KAM theory. We get a much better regularity result in a very simple and natural way.

(3) The key point in the approach is that we used the method of isolating block from [CZ], which was mainly developed by C.C.Conley [C]. It is fascinating to see that, this method perfectly matches our problem, it not only gives us the existence theorem, but also provides us with the regularity estimation at the same time. Here we mean by "regularity" the closeness of the Birkhoff periodic orbits of the perturbation map to the torus for the unperturbed map.

(4) We can not get the uniform estimation as in the convex case for the minimum Birkhoff periodic orbit, see [BK]. We present M.Herman's example [H2] which says that in non-convex situations, oscillations do happen. Thus in these non-convex cases, these "isolating blocks" give optimal estimation.

2. Preliminaries and formulation of result. The general set up is taken from [BK]. We modify some notions and simplify the reduction steps.

Consider the cotangent bundle of n-torus, $\mathbf{T}^n \times \mathbf{R}^n = \{(\phi_1, \cdots, \phi_n, r_1, \cdots, r_n), \phi_i \in \mathbf{R}/\mathbf{Z}, r_i \in \mathbf{R}, i = 1, \cdots, n\}$ with the natural symplectic 2-form:

$$\Omega = \sum_1^n d\phi_i \wedge dr_i$$

Let $f_0 : \mathbf{T}^n \times U \to \mathbf{T}^n \times U$ be an integrable symplectic diffeomorphism, i.e. an Ω-preserving diffeomorphism of the form:

$$f_0(\phi, r) = (\phi + a(r), r), \phi \in \mathbf{T}^n, r \in U$$

*Dept.of Math, Cal.Tech.

Here U is an open set in $\mathbf{R}^n$.

Let furthermore $F_0 : \mathbf{R}^n \times U \to \mathbf{R}^n \times U$ be a lift of f_0 to the universal cover s.t. for $x \in \mathbf{R}^n, r \in U$

$$F_0(x,r) = (x + a(r), r) \tag{0.1}$$

As indicated in the title, we assume the following nondegeneracy condition for the Hessian of the generating function:

(R) $a : U \to \mathbf{R}^n$ is a regular injective map.

Let us explain why (R) is the nondegeneracy condition for the Hessian of the generating function. Recall that the generating function for F_0 is a function $H_0(x,x')$ so that $F_0(x,r) = (x',r')$ if and only if

$$r = \frac{\partial}{\partial x} H_0(x,x'), r' = -\frac{\partial}{\partial x'} H_0(x,x') \tag{0.2}$$

It follows from (0.1) and (0.2) that H_0 depends on the difference $x' - x : H_0(x,x') = h(x'-x)$. Now $r = r' = -Dh(x'-x) = -Dh(a(r))$, or $Dh = -a^{-1}$, therefore, the nondegeneracy condition for the Hessian of the generating function is exactly the regularity condition for the "twist" a.

Condition (R) is the standard condition for the KAM theory.

Properly choose a coordinate system, h can be written in the following normalized form:

$$h(x'-x) = \frac{1}{2}\sum_{i=1}^{s}(x'_i - x_i)^2 - \frac{1}{2}\sum_{i=s+1}^{n}(x'_i - x_i)^2 \tag{0.3}$$

where s is the signature of the Hessian of h.

Theorem A in [BK] is the special case $s = n$.

It should be mentioned that perturbation is the perturbation of the generating function. Let H, F, f be the perturbation of H_0, F_0, f_0 respectively, H is the generating function of F, i.e. $F(x,r) = (x',r')$ if and only if

$$r = \frac{\partial}{\partial x} H(x,x'), r' = -\frac{\partial}{\partial x'} H(x,x')$$

and f preserves the r-component of the center of masses on each torus $\mathbf{T}^n \times \{r_0\}$ for some $r_0 \in U$, or equivalently, for any $m \in \mathbf{Z}^n$,

$$H(x+m, x'+m) = H(x,x') \tag{0.4}$$

Let $r_0 \in U, s_0 = a(r_0)$. We want to find the periodic orbits which are close to the torus $\mathbf{T}^n \times \{r_0\}$. We show the existence of such orbits if the vector s_0 has rational coordinates.

Let $(\phi, r) \in \mathbf{T}^n \times \mathbf{R}^n$ be a periodic orbit of the map f with period q. Let $(x,r) \in \mathbf{R}^n \times \mathbf{R}^n$ be a lift of (ϕ, r), then there exists a vector $\omega \in \mathbf{Z}^n$ s.t. $F^q(x,r) = (x+\omega, r)$. The vector ω/q is called the rotation vector of the point (ϕ, r), it depends on the choice of the lift F but it is uniquely defined modulo $\mathbf{Z}^n$.

We shall prove the following:

THEOREM. *Let f be a perturbation of an integrable symplectic map f_0 satisfying (R), it is defined by the generating function $H = h + P$ where P satisfies (4). Let $\omega = (\omega_1, \cdots, \omega_n) \in \mathbf{Z}^n$, s.t. $\omega_1, \cdots, \omega_n, q$ are relatively prime and the vector $s_0 = \omega/q \in a(U)$. Denote $r_{\omega,q} = a^{-1}(\omega/q)$.*

Conclusion: there exists $\Delta = \Delta(f_0)$ only depends on f_0 but not on ω and q s.t. if $\delta = \| P \|_{c^1} \leq \Delta$, then f has at least $n+1$ different periodic orbits with the rotation vector ω/q which lie completely inside the $C\delta$ neighbourhood of the torus $\mathbf{T}^n \times \{r_{\omega,q}\}$, where $C = C(n,q)$ depends only on the dimension of the manifold and the length of the periodic orbit.

3. Periodic states with the given rotation vector. Fix ω, q as in the theorem, consider the space $\Psi_{\omega,q}$ of "periodic states", where $\Psi_{\omega,q} \ni x = (x^0, \cdots, x^q), x^i \in \mathbf{R}^n$ if and only if it satisfies the following boundary condition:

$$x^q = x^0 + \omega \tag{0.5}$$

Define an action on $\Psi_{\omega,q}$:

$$L_{\omega,q}(x) = \sum_{i=0}^{q-1} H(x^i, x^{i+1}) \tag{0.6}$$

A point $x \in \Psi_{\omega,q}$ is a critical point of the action if

$$0 = \frac{\partial L_{\omega,q}}{\partial x^i} = \frac{\partial H}{\partial x'}(x^{i-1}, x^i) + \frac{\partial H}{\partial x}(x^i, x^{i+1}), \forall i \in \mathbf{Z}$$

It is readily seen that $x \in \Psi_{\omega,q}$ satisfying $x^{i+1} - x^i \in a(U)$ is a critical point if and only if $\{(x^i, r^i) : i = 0, \cdots, q\}$, where $r^i = \frac{\partial H}{\partial x}(x^i, x^{i+1})$, is a periodic orbit for the map F.

We have now reduced our problem to be finding the critical point $x \in \Psi_{\omega,q}$, of the action $L_{\omega,q}, x$ satisfying $x^{i+1} - x^i \in a(U)$.

Remember the condition (0.4), let G be the group generated by the translations $T_m : (x^0, \cdots, x^q) \to (x^0 + m, \cdots, x^q + m), m \in \mathbf{Z}^n$, then $\mathrm{L}_{\omega,q}$ is G-invariant. Hence $L_{\omega,q}$ acts on the quotient space $\Phi^*_{\omega,q} = \Psi_{\omega,q}/G$. Now use the coordinate system introduced by [BK]:

$$v = \frac{1}{q}(x^0 + \cdots + x^{q-1}), t^i = x^i - x^{i-1} - \omega/q, i = 1, \cdots, q-1$$

In terms of this coordinate system, we have

$$T_m(v,t) = (v + m, t), t = (t^1, \cdots, t^{q-1})$$

Hence $\Phi^*_{\omega,q} = \mathbf{T}^n \times \mathbf{R}^{n(q-1)}$, where v is the torus $\mathbf{T}^n$-coordinate and t is the $\mathbf{R}^{n(q-1)}$ coordinate.

4. Proof of the theorem. We consider the simplest case $n = 2, s = 1$. The general case proceeds alone the same line, and we will indicate the crucial part in the proof of the general situation after the proof of this simple case.

When $n = 2, s = 1$, our generating function is

$$h(x, x') = \frac{1}{2}(x'_1 - x_1)^2 - \frac{1}{2}(x'_2 - x_2)^2$$

where $x = (x_1, x_2), x' = (x'_1, x'_2)$.

Let $x^i = (x^i_1, x^i_2), t^i = (t^i_1, t^i_2), v = (v_1, v_2)$, then the action under the new coordinate system (v, t) is

$$L(v,t) = \sum_{i=0}^{q-2}\{\frac{1}{2}(t_1^{i+1} + \omega_1/q)^2 - \frac{1}{2}(t_2^{i+1} + \omega_2/q)^2 + P\}$$
$$+ \frac{1}{2}(t_1^1 + \cdots + t_1^{q-1} - \omega_1/q)^2 - \frac{1}{2}(t_2^1 + \cdots + t_2^{q-1} - \omega_2/q)^2 + P$$

Take the derivative with respect to t^j_i,

$$\begin{array}{ll} \frac{\partial L}{\partial t_1^1} & 2t_1^1 + t_1^2 + \cdots + t_1^{q-1} + R_1^1(dP) \\ \frac{\partial L}{\partial t_1^2} & t_1^1 + 2t_1^2 + \cdots + t_1^{q-1} + R_1^2(dP) \\ \vdots & \vdots \\ \frac{\partial L}{\partial t_1^{q-1}} & t_1^1 + t_1^2 + \cdots + 2t_1^{q-1} + R_1^{q-1}(dP) \end{array}$$

and

$$\begin{array}{ll} \frac{\partial L}{\partial t_2^1} & -2t_2^1 - t_2^2 - \cdots - t_2^{q-1} + R_2^1(dP) \\ \frac{\partial L}{\partial t_2^2} & -t_2^1 - 2t_2^2 - \cdots - t_2^{q-1} + R_2^2(dP) \\ \vdots & \vdots \\ \frac{\partial L}{\partial t_2^{q-1}} & -t_2^1 - t_2^2 - \cdots - 2t_2^{q-1} + R_2^{q-1}(dP) \end{array}$$

where

$$L = L_{\omega,q}$$

and

$$| R_i^j(dP) | \leq 2q\delta$$

Now proceed as in [CZ], consider the gradient flow $\frac{d}{ds}(v,t) = \nabla L(v,t)$, and let $t_i = (t_i^1, \cdots, t_i^{q-1}) \in \mathbf{R}^{q-1}, i = 1, 2$. Write down the gradient equation for (t_1, t_2)-component, we have

$$\frac{d}{ds}\begin{pmatrix} t_1 \\ t_2 \end{pmatrix} = \begin{pmatrix} A & 0 \\ 0 & -A \end{pmatrix}\begin{pmatrix} t_1 \\ t_2 \end{pmatrix} + \begin{pmatrix} R_1(dP) \\ R_2(dP) \end{pmatrix} \tag{0.7}$$

where

$$A = \begin{pmatrix} 2 & 1 & \cdots & 1 \\ 1 & 2 & \cdots & 1 \\ \vdots & & & \\ 1 & 1 & \cdots & 2 \end{pmatrix}$$

is a $(q-1) \times (q-1)$ matrix. The crucial thing we need is that A is non-singular, it has 1 and q as its eigenvalue, and the eigenvalue 1 has multiplicity $q-2$.

Now we arrive at the same situation as the one in [CZ]. We construct the isolating block in the sense of C.C.Conley. Let $D_1 = \{t_1 \in \mathbf{R}^{q-1} :\mid t_1 \mid \leq 3q^2\delta\}, D_2 = \{t_2 \in \mathbf{R}^{q-1} :\mid t_2 \mid \leq 3q^2\delta\}$.

Claim: $B = \mathbf{T}^n \times D_1 \times D_2$ is an isolating block

Proof. for $\mid t_1 \mid \geq 3q^2\delta$,
$\frac{d}{ds}\frac{1}{2} \mid t_1 \mid^2 = < t_1, At_1 + R_1(dP) >$
$\geq \mid t_1 \mid^2 - 2q^2\delta \mid t_1 \mid = \mid t_1 \mid (\mid t_1 \mid - 2q^2\delta) \geq q^2\delta \mid t_1 \mid > 0.$
Similarly, for $\mid t_2 \mid \geq 3q^2\delta, \frac{d}{ds}\frac{1}{2} \mid t_2 \mid^2 \leq -q^2\delta \mid t_2 \mid < 0$. Hence, B is an isolating block, the claim is proved.

Now, Conley-Zehnder argument [CZ,Thm4] concludes that there is an invariant set Σ for the gradient flow, lying **inside** the isolating block B. the cup length satisfies:

$$l(\Sigma) \geq n+1$$

To get the $n+1$ geometrically different periodic orbits for our perturbation map, we need to module out the shift map $S : \Phi^*_{\omega,q} \to \Phi^*_{\omega,q}, (x^0, x^1, ..., x^{q-1}, x^q) \to (x^1, x^2, ..., x^q, x^1 + \omega)$, since L is invariant under S. We use the following argument by Golé[G].Under the (v,t)-coordinates, it takes the following form:

$$S(v, t^1, t^2, ..., t^{q-1}) = (v + \omega/q, t^2, ..., t^{q-1}, -\sum_{i=1}^{q-1} t^i)$$

Now $S = S_1 \times S_2$, where $S_1 : \mathbf{T}^n \to \mathbf{T}^n, v \to v + \omega/q$ is a q-periodic, fixed point free diffeomorphism, and $S_2 : \mathbf{R}^{n(q-1)} \to \mathbf{R}^{n(q-1)}, (t^1, ..., t^{q-1}) \to (t^2, ...t^{q-1}, -\sum_{i=1}^{q-1} t^i)$ is a linear isomorphism. $\Phi^*_{\omega,q}$ is a q-fold covering of $\Phi_{\omega,q} = \Phi_{\omega,q}/\{S^i\}_{i=0}^{q-1}$, the latter is a fiber bundle over $\mathbf{T}^n$, which also has the homotopy type of $\mathbf{T}^n$. Now we have the following covering maps π and deformation retracts k_1, k_2:

$$\begin{array}{ccc} \Phi^*_{\omega,q} = \Psi_{\omega,q}/G & \longrightarrow & \mathbf{T}^n \\ \pi \big\downarrow & & \\ \Phi_{\omega,q} = \Phi^*_{\omega,q}/\{S^i\}_{i=0}^{q-1} & \longrightarrow & \mathbf{T}^n \end{array}$$

Let $K^* = k_2^* \circ (k_1^*)^{-1} : H^*(\Phi^*_{\omega,q}) \longrightarrow H^*(\mathrm{T}^n) \longrightarrow H^*(\Phi_{\omega,q})$. Then K^* is an isomorphism on cohomology. Where H^* is the Alexander cohomology with real coefficient.

L induces an action on $\Phi_{\omega,q}$ which will still be denoted by L. Now the critical points of L on $\Phi_{\omega,q}$ are in one-to-one correspondence with the geometrically different periodic orbits with the rotation vector ω/q of F.

The existence result in the theorem follows from the following lemma.

Lemma. Let $\underline{\Sigma} = \pi(\Sigma)$ be the invariant set for the gradient flow in $\underline{B} = \pi(B)$. Then

$$l(\underline{\Sigma}) \geq l(\Sigma) \geq n+1$$

Proof. We have the following diagram ($\Phi^* = \Phi^*_{\omega,q}, \Phi = \Phi_{\omega,q}$):

$$\begin{array}{ccccc}
H^n(\Phi) & \xrightarrow{i_1^*} & H^n(\underline{B}) & \xrightarrow{j_1^*} & H^n(\underline{\Sigma}) \\
K^* \uparrow \downarrow \pi^* & & \downarrow \pi^* & & \downarrow \pi^* \\
H^n(\Phi^*) & \xrightarrow{i_2^*} & H^n(B) & \xrightarrow{j_2^*} & H^n(\Sigma)
\end{array}$$

where i_1, i_2, j_1, j_2 are inclusions. We need the following facts:

(1) i_2^* is an isomorphism: B is a deformation retract of Φ^* (see [CZ]).

(2) $\pi^* \circ K^*$ is the multiplication by $d \neq 0$ in $R \cong H^n(\Phi^*)$.

(3) Forget K^*, the above diagram commutes.

(4) $j_2^*(w^*) \neq 0$, where $w^* = w_1^* \cup ... \cup w_n^*$, where w_i^*'s generate $H^1(B)$. This is the core of [CZ, Thm4].

Take $w = K^* \circ (i_2^*)^{-1}(w^*) = w_1 \cup ... \cup w_n$, where $w_i = K^* \circ (i_2^*)^{-1} w_i^*$, then w generates $H^n(\Phi)$. Then $j_1^* o i_1^*(w) \neq 0$, otherwise $0 = \pi^* o j_1^* o i_1^*(w) = j_2^* o i_2^* o \pi^*(w) = j_2^* \circ i_2^* \circ \pi^*(K^* \circ (i_2^*)^{-1}(w^*)) = j_2^* \circ i_2^* \circ (\pi^* \circ K^*) \circ (i_2^*)^{-1}(w^*) = d j_2^* \circ i_2^* \circ (i_2^*)^{-1}(w^*) = d j_2^*(w^*)$, a contradiction. Hence

$$(j_1^* \circ i_1^* w_1) \cup ... \cup (j_1^* \circ i_1^* w_n) \neq 0$$

and thus, $l(\underline{\Sigma}) \geq l(\Sigma) \geq n+1$. The lemma is proved.

Now Conley-Zehnder argument concludes that L has at least $n+1$ critical points **inside** the isolating block. It is left to show that, corresponding to any of those critical points, say $x, x^i - x^{i-1} \in a(U), \forall i \in \mathbf{Z}$. We prove this together with the regularity as follows:

Since the critical points found all lie inside B, we get

$$\mid t_j^i \mid \leq 3q^2\delta \tag{0.8}$$

for all i, j, but $t^i = x^i - x^{i-1} - \omega/q$, above is exactly the regularity result. And it is also from this regularity, we can assure that $x^i - x^{i-1} \in a(U)$ for δ small enough, because $\omega/q = a(r_{\omega,q}), r_{\omega,q} \in U$, and $a(U)$ is open.

5. Remarks for the general situations. The simplest case $n = 2, s = 1$ is proved. For the general case it can be computed that, neglecting the torus coordinate, the gradient of the action looks like

$$\frac{d}{dS}\begin{pmatrix} t_1 \\ t_2 \\ \vdots \\ t_n \end{pmatrix} = \begin{pmatrix} A & & & & & 0 \\ & \ddots & & & & \\ & & A & & & \\ & & & -A & & \\ & & & & \ddots & \\ 0 & & & & & -A \end{pmatrix} \begin{pmatrix} t_1 \\ t_2 \\ \vdots \\ t_n \end{pmatrix} + \begin{pmatrix} R_1(dP) \\ R_2(dP) \\ \vdots \\ R_n(dP) \end{pmatrix}$$

where the number of A's is the signature s of the Hessian.This big matrix is in block diagonal form, it is nonsingular. Hence Conley-Zehnder method works in these general cases.

6. Herman's example. Consider the integrable map $L_B : T^*(\mathbf{T}^2) \to T^*(\mathbf{T}^2)$

$$L_B(\theta, r) = (\theta + rB, r)$$

where

$$B = \begin{pmatrix} 0 & 1 \\ 1 & 0 \end{pmatrix}$$

$\theta = (\theta_1, \theta_2), r = (r_1, r_2)$. Note that B has eigenvalue $1, -1$.

To construct the small perturbation, let $\phi(\theta_1, \theta_2) = \frac{\delta}{2\pi} sin(2\pi\theta_1)$ be a function on $\mathbf{T}^2$, and $G_\phi(\theta, r) = (\theta, r + \frac{\partial \phi}{\partial \theta}) = (\theta_1, \theta_2, r_1 + \phi'(\theta_1), r_2)$. Finally, let $F = G_\phi \circ L_B$.

Write down the lift map for F (still denoted by F) and its iterates explicitely:

$$F(x, r) = (x_1 + r_2, x_2 + r_1, r_1 + \delta cos(2\pi(x_1 + r_2)), r_2) \tag{0.9}$$

For $j \geq 2$, the j-th iterate of F is

$$\begin{aligned} F^j(x, r) &= (x_1^j, x_2^j, r_1^j, r_2^j) \\ &= (x_1 + jr_2, x_2 + jr_1 + \delta \sum_{k=1}^{j-1} (j-k) cos(2\pi(x_1 + kr_2)), \\ &\qquad r_1 + \delta \sum_{k=1}^{j} cos(2\pi(x_1 + kr_2)), r_2) \end{aligned}$$

Take any integer $q \geq 2$, and an integer vector $\omega = (1, 1)$. Then the unperturbed torus with rotation vector ω/q is $r = r_{\omega,q} = (1/q, 1/q)$. Through an elementary calculation, we find that $\{F^i(-1/(2q), x_2, 1/q, 1/q), i \in \mathbf{Z}\}$ is a Birkhoff periodic orbit with rotation vector ω/q, where x_2 can be arbitrary. The closeness of the Birkhoff periodic orbit to the unperturbed torus is measured by $x^{j+1} - x^j - \omega/q$, in this example $x^{j+1} - x^j - \omega/q = (0, r_1^j - r_1)$. For $1 \leq j \leq q$,

$$\begin{aligned} r_1^j - r_1 &= \delta \sum_{k=1}^{j} cos(2\pi(k - 1/2)r_2) \\ &= \delta sin(2\pi j r_2)/(2 sin(\pi r_2)) \end{aligned}$$

where $r_2 = 1/q$. Take $j = [q/4]$ the integer part of $q/4$, for large q, we have

$$| r_1^j - r_1 | \approx \frac{\delta q}{2\pi}$$

In [prop.2 BK], they can get uniform estimation for the minimum Birkhoff periodic orbit, i.e, the right hand side of the above estimation does not depends on q. Now this example tells that in the non-convex situations, we can not get the uniform estimation, roughly speaking, (0.8) is optimal.

Acknowledgement. This work is done under the direction of my advisor A.Katok. I want to thank Professor M.Herman for telling me his example. I also want to thank Dr.Christophe Golé for pointing out an error in my preprint and for giving me a proof to fix it.

REFERENCES

[BK] D.Bernstein and A.Katok, *Birkhoff periodic orbit for small perturbations of completely integrable Hamiltonian system with convex Hamiltonian*, Invent. Math., 88 (1987), pp. 225–241.

[C] C.C.Conley, *Isolated invariant set and Morse index*, CBMS, Reginal conf. in Math., 38 (1978).

[CZ] C.C.Conley and E.Zehnder, *The Birkhoff-Lewis fixed point theorem and a conjecture of V.I. Arnold*, Invent. Math., 73 (1983), pp. 33–49.

[G] Christophe Golé, *Thesis*.

[H1] M.Herman, *Existence et non existence de tores invariants par des diffeomorphsmes symplectiques*, preprint (1988).

[H2] M.Herman, Oral communication.

[K] A.Katok, *Minimal orbits for small perturbations of completely integrable Hamiltonian systems*, preprint.

[M] M.Muldoon, *Ghosts of order in Hamiltonian systems with many degree of freedom*, Ph.D. thesis, Caltech (1988).

PHYSICAL EXAMPLES OF LINKED TWIST MAPS WITH CHAOTIC DYNAMICS

VICTOR J. DONNAY*

0. Introduction. In [D1][D2] [D3][DL], we constructed new examples of chaotic dynamical systems arising from "physical" systems: billiards, particle motion in a potential field, and geodesic flows on a surface. By chaotic we mean that these systems have positive Lyapunov exponents almost everywhere; hence by the general Pesin theory [P] they have positive measure theoretic entropy, ergodic components of positive measure and on the ergodic components the system exhibits very strong stochastic behaviour (essentially Bernoulli). In many cases, the systems have also been shown to be ergodic.

One can interprete the dynamics in these systems as being generated by a composition of twist maps. Such systems are called linked twist maps and have been studied in an abstract setting (see [BE][W1] and references therein). Thus our work shows that chaotic linked twist maps can arise from physical systems. In the abstract setting, the individual twist maps that were composed to form the linked twist map system were integrable. We generalize this result to show that non-integrable twist maps can also be used.

Wojtkowski [W2] introduced a method to prove positive Lyapunov exponents that uses cone-fields. This method is particularly well suited for analyzing linked twist maps and gives a conceptually simple proof of positivity of Lyapunov exponents. For certain classes of smooth systems, Burns and Gerber [BG] (see also [K]) have shown that the cone-field method can be used to prove ergodicity.

We will describe how our various systems can be considered as linked twists maps and outline the proof that they have positive Lyapunov exponents. Full details can be found in the original papers. Using this approach, we construct a new example of a metric on the two-torus for which the geodesic flow is ergodic (see §11).

1. Twist maps and cone-fields. In what follows, we will deal with a compact two-dimensional Riemannian manifold M, a map $\Phi : M \to M$ and a smooth invariant measure μ. In the simplest case, M will be the annulus, $M = \{(s,\theta) | 0 \leq s < 2\pi, \, 0 \leq \theta < \pi\}$.

In the case of an annulus, Φ is a twist map if for $(s_1, \theta_1) = \Phi(s,\theta)$ we have that

$$\frac{\partial s_1}{\partial \theta} > 0, \tag{1}$$

i.e. the image of a vertical curve in the domain has monotonically increasing s coordinate.

*Department of Mathematics, Bryn Mawr College, Bryn Mawr, Pa. 19010

For our systems, the maximal Lyapunov exponents λ^+, defined by

$$\lambda_+(z) = \lim_{n\to\infty} \frac{1}{n} \ln ||D\Phi^n(z)||,$$

exist for almost every $z \in M$ [O][KS]. Wojtkowski's [W2] criterion for showing that λ_+ is positive involves cones of vectors in the tangent space. A cone $\mathcal{C}(z)$ in the tangent space T_zM at $z \in M$ is a subset $\mathcal{C} = \{aX_1 + bX_2 : ab \geq 0\} \subset T_zM$ where X_1 and X_2 are linearly independent vectors and $a, b \in R$. We call Interior $(\mathcal{C}(z)) = \{aX_1 + bX_2 : ab > 0 \text{ or } a = b = 0\}$ the interior of $\mathcal{C}$. A measurable (continuous) cone-field for our space M is a family of cones $\{\mathcal{C}(z)\} \subset T_zM$ defined for μ almost every $z \in M$ such that the vectors $X_1(z), X_2(z)$ vary measurably (continuously) with z.

THEOREM 1 [W2]. *Let $\mathcal{C}(z)$ be a measurable cone-field such that for almost every z*

$$D\Phi\,(\mathcal{C}(z)) \subset \mathcal{C}(\Phi z), \tag{2}$$

and for almost every z there exists $k(z)$ for which

$$D\Phi^{k(z)}\,(\mathcal{C}(z)) \subset \text{ Interior}\, \mathcal{C}(\Phi^{k(z)} z). \tag{3}$$

Then the Lyapunov exponents $\lambda_+(z)$ are positive for almost every $z \in M$.

If (2) holds, we speak of the cone-field as being invariant and if (3) also holds the cone-field is (eventually) strictly invariant.

If Φ were smooth and the cone at each point got squeezed inside the next cone by a uniform amount, then the invariance of the measure would force vectors in the cone to expand uniformly, thereby producing a uniformly hyperbolic system (Fig.1) Under the weaker hypothesis that it takes a certain number of iterations before a cone gets squeezed inside the image cone and that the amount of squeezing is arbitrarily small, Wojtkowski used the ergodic theorem to show that on average almost every point will have tangent vectors that expand at an exponential, but non-uniform, rate. This method has the advantage that it does not require any estimates of expansion rates. Instead, it only needs that expanding directions at different points are aligned in compatible ways.

The Burns-Gerber result [BG] can be phrased as follows:

THEOREM 2. *Let Φ be smooth and $z \in M$ a point for which (2,3) hold. If the cone-field is continuous at z, then there is an open neighborhood of z that belongs (modulo a set of zero measure) to one ergodic component.*

One can prove global ergodicity by showing that the set of z for which this theorem applies is of full measure and is connected. If Φ has singularities, the method of Sinai and Chernov[SC][KSS] can often be used to prove ergodicity.

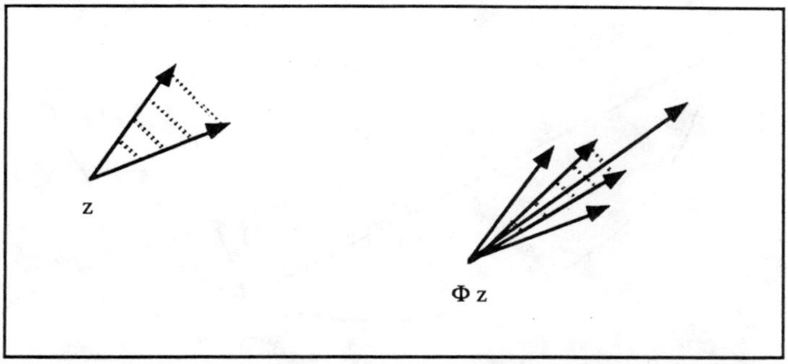

Fig. 1. Uniform squeezing of cones produces expansion

2. Billiards. We describe how these notions apply in the case of billiards. First some background [CFS].

Let Q be a connected domain in the plane with piecewise smooth boundary. By a billiard in Q we mean the dynamical system arising from the uniform motion of a point mass inside Q with elastic reflections at the boundary: the angle of reflection equals the angle of incidence. This motion produces a flow ψ^t in the space T_1Q of unit tangent vectors of Q with the obvious identifications. Let $\pi_1 : T_1Q \to Q$ denote the natural projection.

Let $\Phi : M \to M$ be the standard section map, where $M \subset T_1Q$ is the two dimensional manifold consisting of unit vectors attached at the boundary ∂Q and pointing inside Q. For $z \in M$, the point Φz is gotten by following the point z under the billiard flow ψ^t until its next collision with the boundary.

On M, we introduce coordinates (s, θ) where s is the arc length parameter along ∂Q and $\theta \in [0, \pi]$ is the angle which the unit vector makes with ∂Q (Fig. 2). We denote by $\{X_s, X_\theta\}$ the associated tangent vectors. Φ preserves the measure $\mu = c \sin\theta \, ds \, d\theta$, where c is a normalizing constant. The points in ∂Q at which the piecewise smooth boundary components meet will lead to singularities for Φ.

Remark 3: The billiard map Φ for a convex domain is a twist map. If we take the vertical family $\eta(\sigma) = (s, \theta + \sigma)$, then the image $\Phi\eta(\sigma) = (s_1(\sigma), \theta_1(\sigma))$ satisfies $\frac{ds_1}{d\sigma} > 0$.

The first billiard for which chaotic behaviour was proven was the Sinai scatter [S] in which one has a circular (or more generally concave) obstacle on the torus. The intuitive idea for the chaotic behaviour was that a diverging family of trajectories was made to diverge even more strongly by collision with the concave obstacle. Thus the behaviour of trajectories was similar to the case of geodesic flow on surfaces of negative curvature. Hopf [H] had used this property of divergence to prove ergodicity of geodesic flow on surfaces of negative curvature.

For a region with a convex boundary, collision with the boundary causes a diverging family of trajectories to converge rather than diverge. Thus one did not

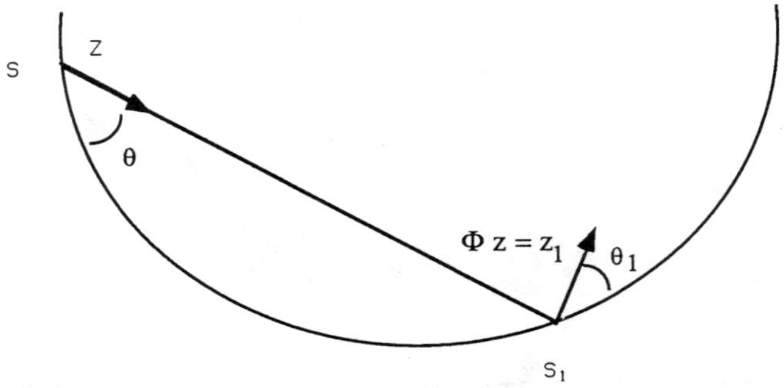

Fig. 2. Billiard Map

expect that chaotic behaviour could be produced in a convex billiard. This notion was further strengthened by the work of Lazutkin [L] who should that a smooth, strictly convex billiard will always have caustics and hence can not be ergodic. A caustic is a closed curve in the configuration space to which a trajectory is repeatedly tangent and correspond to invariant curves in the phase space.

Thus one was very surprised by the ergodicity of Bunimovich's stadium billiard [B1][B2]: a convex region consisting of two half-circles joined by straight lines. Note that this billiard is neither smooth nor strictly convex so it does not contradict Lazutkin's results.

By viewing the stadium as a (non-smooth) perturbation of the circle, we can use the cone-field method and the notion of linked twist maps to give a "one-line" proof that the stadium billiard has positive Lyapunov exponents almost everywhere.

For billiards inside a circle of radius r, the billiard map $\Phi(s,\theta) = (s_1,\theta_1)$ is given by

$$\begin{aligned} s_1 &= s + 2r\theta, \\ \theta_1 &= \theta. \end{aligned} \tag{4}$$

This system is integrable: the angle θ stays constant along an orbit. The phase space decomposes into a union of invariant circles given by θ =constant.

We define a cone-field $\{\mathcal{C}(z)\}, z \in M$, by

$$\mathcal{C}(z) = \{aX_s + bX_\theta : ab \geq 0\}. \tag{5}$$

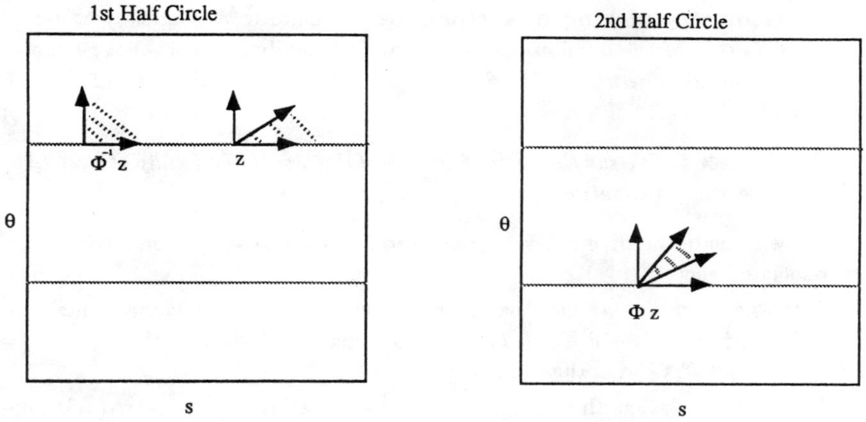

Fig.3. Strict invariance of cones in stadium

Under the map (4), this cone-field is invariant but not strictly invariant (i.e. it satisfies (2) but not (3)):

(6a) $$D\Phi_z(X_\theta) \subset \text{Interior}\,(\mathcal{C}(\Phi z))$$

(6b) $$D\Phi_z(X_s) = X_s \in \partial\,(\mathcal{C}(\Phi z))\,.$$

The first result holds because the billiard map Φ is a twist map (Remark 3). We interpret (6b) as saying that the horizontal vector X_s stays horizontal. In terms of the invariant circles, the tangent to the invariant circle gets sent to the tangent to the invariant circle.

To produce positive Lyapunov exponents, we must find a way to push up the horizontal edge of the cone. We achieve this by perturbing the circle in a non-smooth way to produce the stadium: we cut the circle in half, pull the two halves apart and connect them by straight lines. When a trajectory goes from one half-circle to the other, the horizontal edge of the cone will get "kicked up", so that the cone-field becomes strictly invariant (Fig. 3).

When a trajectory passes from one half-circle to the next, it moves in a straight line in a flat region (zero curvature). As we will explain, such a motion can itself be viewed as an integrable twist map. Thus the stadium billiard is a linked twist map system: a composition of the integrable twist map coming from the billiard motion in the half-circle with the integrable twist map coming from straight line motion in the plane. The twist produced by the second twist map is oriented along a different direction than the twist of the first map and hence serves to push the bottom edge of the cones strictly inside the image cone.

To prove our assertion that in going from one half-circle to the other the horizontal edge of the cone gets pushed up, we look at the tangent vectors geometrically.

3. Geometry of tangent vectors. We associate to a tangent vector $\xi = s'X_s + \theta' X_\theta \in T_zM$ the infinitesimal one-parameter family of points that generates ξ: i.e. the variation $\eta_\xi(\sigma) = (s(\sigma), \theta(\sigma)) = (s + \sigma s', \theta + \sigma\theta'), \sigma \in (-\epsilon, \epsilon)$ for which $\eta_\xi(0) = z,\ \eta'_\xi(0) = \xi$.

DEFINITION 4. *We say the vector ξ focuses if the associated family of rays $\eta_\xi(\sigma)$ focus in linear approximation.*

We will relate the time it takes a family to focus with the slope $m(\xi) = \theta'/s'$ of its associated vector.

Let z_0, $z_1 = \Phi z_0 \in M$ have basepoints on convex boundary components. For $\xi_0 \in T_{z_0}M$, $\bar{\xi} = \bar{s}'X_s + \bar{\theta}'X_\theta \in T_{z_1}M$, we compare the slope of the vector $\xi = D\Phi\xi_0 = s'X_s + \theta'X_\theta$ with the slope of $\bar{\xi}$.

Denote by L the length of the line segment connecting $\pi_1(z_0)$ and $\pi_1(z_1)$. Suppose that under the flow ψ^t the vector ξ_0 focuses at a point on this line segment a distance $d_0(\xi_0)$ from $\pi_1(z_0)$. Also suppose that under the backwards flow ψ^{-t} the vector $\bar{\xi}$ focuses a distance $d_1(\bar{\xi}) > 0$ from $\pi_1(z_1)$.

FOCUSING LEMMA 5: *If*

$$L - d_0(\xi_0) - d_1(\bar{\xi}) > 0 \tag{7}$$

then

$$\frac{\theta'}{s'} > \frac{\bar{\theta}'}{\bar{s}'} \ . \tag{8}$$

Proof. Let

$$\eta(\sigma) = \left(s_1 + \sigma, \theta(\sigma) = \theta_1 + \frac{\theta'}{s'}\sigma\right),$$

$$\bar{\eta}(\sigma) = \left(s_1 + \sigma, \bar{\theta}(\sigma) = \theta_1 + \frac{\bar{\theta}'}{\bar{s}'}\sigma\right)$$

be the families corresponding to ξ and $\bar{\xi}$.

We interpret (7) to mean that under ψ^{-t}, the family $\eta(\sigma)$ focuses further from $\pi_1(z_1)$ than does the family $\bar{\eta}(\sigma)$. Therefore, for a fixed value of σ, one has that

$$\theta(\sigma) > \bar{\theta}(\sigma),$$

which proves (8) (Fig. 4).

□

Returning to the stadium, we examine a point z which lies in one half-circle and whose image point $z_1 = \Phi z$ lies in the other half-circle. We compare the slopes of the vectors $D\Phi_z(X_s), X_s \in T_{z_1}M$ and intend to show that

$$m(D\Phi_z(X_s)) > m(X_s). \tag{9}$$

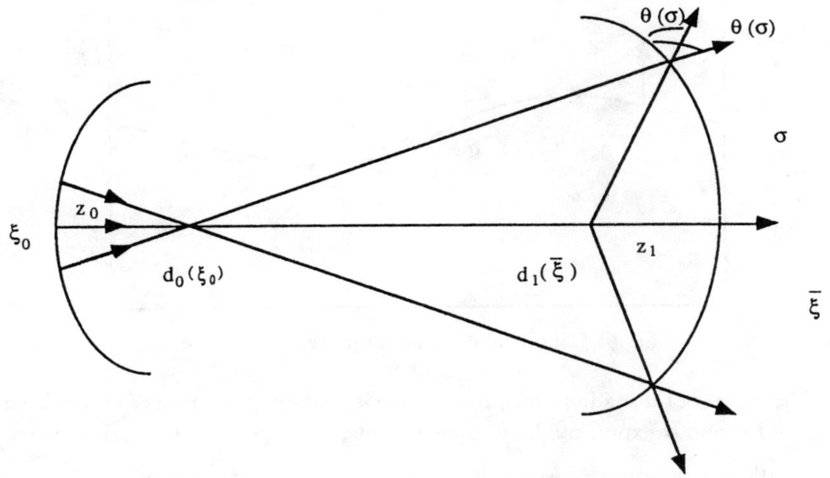

Fig.4. Order of focusing

Let $\eta(\sigma) = (s + \sigma, \theta), \sigma \in (-\epsilon, \epsilon)$ be the family that generates the horizontal vector X_s . In the case of a circle, this family gets sent by (4) to the horizontal vector at (s_1, θ_1). The line segment connecting (s, θ) and (s_1, θ_1) has length $2r \sin\theta$ and by symmetry the family focuses at the mid-point of this line segment.

Hence in the stadium, the vector $X_s \in T_z M$ $(X_s \in T_{z_1} M)$ will focus at a forward (backward) distance $d = r\sin\theta$ $(d_1 = r\sin\theta_1)$. If L is the length of the segment connecting z and z_1, then we claim that

$$L > d + d_1, \tag{10}$$

Note that $2d$ $(2d_1)$ is the length of that part of the trajectory that lies inside the osculating circle at $\pi_1(z)$ $(\pi_1(z_1))$. By pulling the two half circles apart, we insure that both osculating circles lie completely inside the billiard table. Hence $d + d_1 \le \max\{2d_0, 2d_1\} < L$.

Given (10), (9) now follows from the Focusing Lemma 5.

The idea of using the focusing distance to define cones was introduced in [W3].

Remark 6. Along an invariant curve of a twist map there exists a canonical cone-field that is invariant but not strictly invariant. The edges of this cone are the vertical vector and the vector tangent to the invariant curve. The twist nature of the map insures that the vertical vector will get sheared inside the cone under one iteration while the tangent to the invariant curve gets sent to the tangent to the invariant curve (Fig. 5).

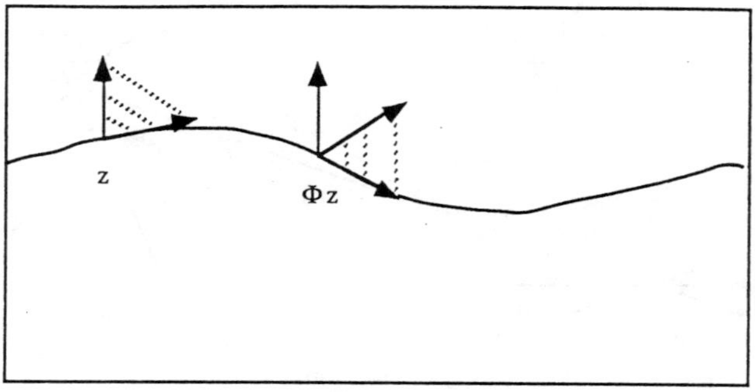

Fig.5. Canonical cone field on invariant curve

The moral of the stadium billiard example is that one can produce systems with positive Lyapunov exponents by composing twist maps that have invariant curves.

4. Focusing arcs. In [D3] we generalize this proof to show that there exists a large class of convex arcs for which the resulting billiard has positive Lyapunov exponents almost everywhere. This class is open in the C^6 topology on curves which allows us to prove that C^6 small perturbations of the Bunimovich stadium billiard have positive Lyapunov exponents. Our examples generalize work of Bunimovich [B1], Wojtkowski [W3], and Markarian [Mr1][Mr2] who had described various non-open classes of arcs.

We say a non-closed curve is convex if when we connect the ends of the curve by a straight line, the resulting closed curve has no self-intersections and bounds a convex set. Henceforth, the term convex curve will signify a C^∞ smooth curve that is non-closed and whose curvature is never zero , i.e. the curve is strictly convex.

We denote by $X_p^-(z) \in T_zM$ the infinitesimal family that under the flow ψ^t is parallel when it reaches z.

Let γ be a convex curve. We examine a ray that collides with γ a finite number of times (Fig. 6).

DEFINITION 7. *An incoming ray is focused by γ if it collides a finite number of times with γ and if the infinitesimal family of rays that is parallel to the incoming ray*

(1) *focuses between each pair of collisions with γ,*

(2) *focuses after hitting γ for the last time.*

For such a ray, the time the parallel family takes to focus after hitting γ for the last time is called its focusing time.

We want boundary curves which do not trap rays: i.e. every (or almost every) ray starting on the boundary will have a finite number of collisions with the curve and then leave it. We make the following restriction:

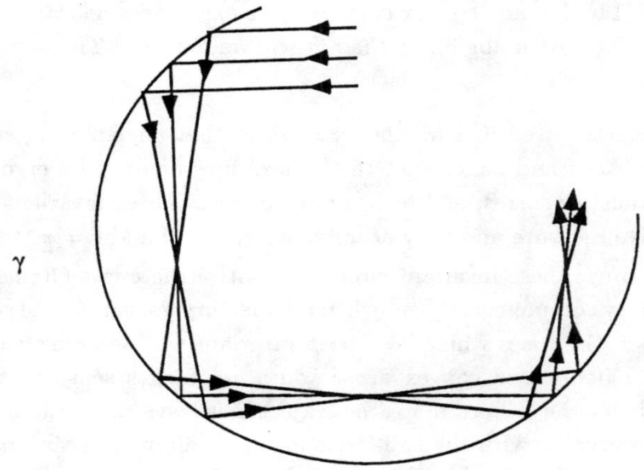

Fig.6. Parallel variation focusing

DEFINITION 8. *A convex arc γ is a convex curve whose curvature $k_\gamma(s)$, $s \in [0, l]$ satisfies*

$$\int_0^l k_\gamma(s)\,ds \le \pi. \tag{11}$$

where the arc-length of γ is l.

DEFINITION 9. *A convex arc γ is focusing if all rays collide with γ only a finite number of times and all rays are focused.*

Note that any sub-arc of a focusing arc is itself a focusing arc.

After our discussion of the stadium billiard, the reader will not be too surprised by the following theorem.

THEOREM 10. *Let Q be a convex region consisting of focusing arcs and straight lines. Suppose that the focusing times for all trajectories that hit the focusing arcs are bounded. Then if the straight edges are sufficiently long, the billiard in Q will have positive Lyapunov exponent almost everywhere.*

Ergodicity of these examples is conjectured but has not yet been definitively proven.

A natural question to ask is under what circumstances will a ray be focusing? And even if a ray is focusing, how can one insure that as $\theta \to 0$, the focusing times remain bounded?

Our principal result is

THEOREM 11. *For any convex curve γ there exists an angle $\Theta(\gamma) > 0$ such that any ray that hits γ with angle less than Θ will be focused. The focusing times for these rays is bounded.*

In [L], Lazutkin used K.A.M. theory to show that any smooth, strictly convex billiard has caustics and hence that the billiard flow can not be ergodic. In phase space, these caustics correspond to invariant circles. These invariant circles fill up a set of positive measure and they accumulate at $\theta = 0$ and $\theta = \pi$.

We wish to use these invariant circles to control the cones (Remark 6) in each convex boundary component. Then, when a trajectory goes from one convex component to another, the cones which have been invariant will become strictly invariant. Actually, for a non-closed convex arc, it does not makes sense to talk about an invariant circle for the billiard map: the trajectories leave the arc and hence are not recurrent. However, as with the half-circle in the stadium, one can image a "ghost" of the invariant circle existing: this is a subset of the invariant circle that would have existed if the arc were part of a closed convex curve. For our purposes, we do not really need an invariant curve; just the possibility that such a curve could exist.

For trajectories that hit the boundary with small angles, Lazutkin [L] introduced the following change of coordinates:

$$
\begin{aligned}
x(s,\theta) &= C_1 \int_0^s r^{-2/3}(s)\,ds, \\
y(s,\theta) &= C_2 r^{1/3}(s)\sin(\theta/2),
\end{aligned}
\tag{12}
$$

where

$$C_1 = \Big(\int_0^l r^{-2/3}(s)\,ds\Big)^{-1},\ l = \text{ length of } \gamma,\ C_2 = 4C_1.$$

Under this change of coordinates the billiard transformation $\Phi : (s,\theta) \mapsto (s_1,\theta_1)$ becomes $(x_1,y_1) = \Phi(x,y)$ where

$$
\begin{aligned}
x_1 &= x + y + y^3 f(x,y), \\
y_1 &= y + y^4 g(x,y).
\end{aligned}
\tag{13}
$$

For small y (equivalently small θ), we see that the billiard map is almost an integrable map. In going from first entering a convex arc to leaving an arc, the horizontal vector X_x will not necessarily stay horizontal, as in the case of the half-circle, but it will not drop very much. The composite map from entering to leaving a convex arc remains a twist map (for all sufficiently small angles). In going from one convex arc to another, the motion in the flat region can still "kick up" this vector.

More quantitatively, if $n(z)$ is the number of hits the point $z = (x,y)$ has with the convex arc then using equation (13) we can show that

$$\lim_{y\to 0} m(D\Phi^{n(z)}X_x) = 0, \tag{14}$$

where $m(\xi)$ is the slope of the vector ξ in the (x,y) coordinates.

Translating (14) into a statement about configuration space, one finds that the parallel variation X_p^- that starts at z will be converging when it leaves $\Phi^k z$, $k = 0, 1, 2, \ldots, n(z)$. Hence it must focus between $\Phi^k z$ and $\Phi^{k+1} z, k = 0, 1, 2, \ldots, n(z) - 1$. This allows us to prove Theorem 11.

A corollary is

THEOREM 12. *Let γ be any convex curve and q any point on γ. Then there exits a neighborhood $U(q) \subset \gamma$ of q such that the convex arc $U(q)$ is focusing.*

The set of focusing arcs form an open set in the space of smooth, convex curves satisfying (11).

THEOREM 13. *Let γ be a focusing arc. If $\tilde{\gamma}$ is a convex arc of the same length that is sufficiently close to γ in the C^6 topology, then $\tilde{\gamma}$ is also a focusing arc.*

The reason that we need to use C^6 is that we made such a good change of variables in (12) that the remainder terms in (13) depend on the fourth derivative of curvature and hence on the sixth derivative of the curve γ. Other authors [Mr2] [B3] have used a weaker change of coordinates to show that the C^6 requirement can by replaced by C^4. Evidence of our "overkill" is that the term in (14) goes to zero like y^3.

If γ is chosen to be a half circle, then Theorem 13 implies that perturbations of the stadium billiard have positive Lyapunov exponents. Fixing the length of the flat sides determines how close to a half circle $\tilde{\gamma}$ must be. Note that the resulting boundary will be, at best, only C^1 smooth.

By Theorem 13, one can replace the half-circles of the stadium by half-ellipses cut along their semi-minor axis and with eccentricity close to 1. We can show [D3 §7] that a half-ellipse with eccentricity significantly different from one is also a focusing arc. This allows us to construct what we call an elliptical stadium (two half-ellipses joined by straight lines) for which the billiard will have positive Lyapunov exponent almost everywhere. We do not have an explicit estimate on how long the flat sides of the stadium must be.

THEOREM 14. *The half-ellipse $\frac{x^2}{a^2} + \frac{y^2}{b^2} = 1$, $x \geq 0$, is a focusing arc if and only if $a/b < \sqrt{2}$.*

5. Model problem. The examples arising from potential fields and geodesic flow be considered as special cases of a certain model problem. This model problem is a composition of integrable twist maps.

We start with a two-dimensional torus T^2 on which we specify a disk. The flow ψ^t occurs in the unit tangent space of the torus and preserves a smooth invariant measure. Given a point and a direction at that point, ψ^t is straight line motion (i.e. geodesic flow on the flat torus) as long as the point remains outside the disk. This motion corresponds to an integrable twist map.

When a particle enters the disk, it will be rotated around the disk a certain amount and then leave the disk. We give the disk D polar coordinates (r, θ), $r \in [0, R]$, $\theta \in [0, 2\pi)$ and the boundary of the disk ∂D coordinates (θ, ϕ) with $\phi \in$

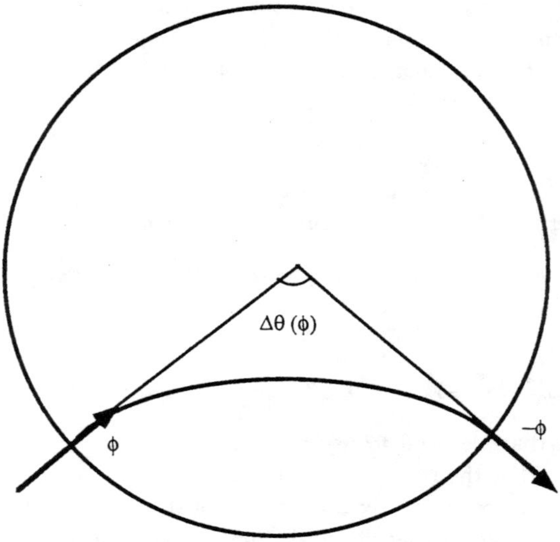

Fig.7. Rotation function $\Delta\Theta(\phi)$

$(-\pi, \pi]$. If a particle enters the disk at a point $\theta, \theta \in [0, 2\pi]$, with angle $\phi \in [0, \pi]$, then the particle will leave the disk at the point $\theta + \Delta\theta(\phi)$ with angle $-\phi$ (Fig. 7). The function $\Delta\theta(\phi)$ is called the **rotation function** and determines the stochastic properties of the flow.

The map from entering to leaving the disk, given by

$$(15) \qquad (\theta, \phi) \to (\theta + \Delta\theta(\phi), -\phi), \quad \phi \in [0, \pi],$$

is an integrable map which in many cases will also be a twist map.

We will define cones for all points that lie outside the disk and derive conditions on the rotation function that will insure that the cone-field is strictly invariant (2,3) and hence that the system has positive Lyapunov exponents are positive almost everywhere. The analysis, although done for the case of one disk on the torus, can easily extended to the case of several disks (see §11).

6. Geodesic flow on the flat torus. Given an orientable surface with a Riemannian metric, the geodesics are the paths on the surface that locally give the shortest distance between points. These geodesics determine a flow ψ^t on the unit tangent bundle of the surface in the following manner. Start at a point on the surface and a unit tangent vector at that point. Flow along the geodesic in the direction of this vector for a distance t. One ends up at a new point on the surface pointing in the direction of a new unit tangent vector. For the flat torus, the geodesics are straight lines. The geodesic flow is integrable; the direction of motion stays

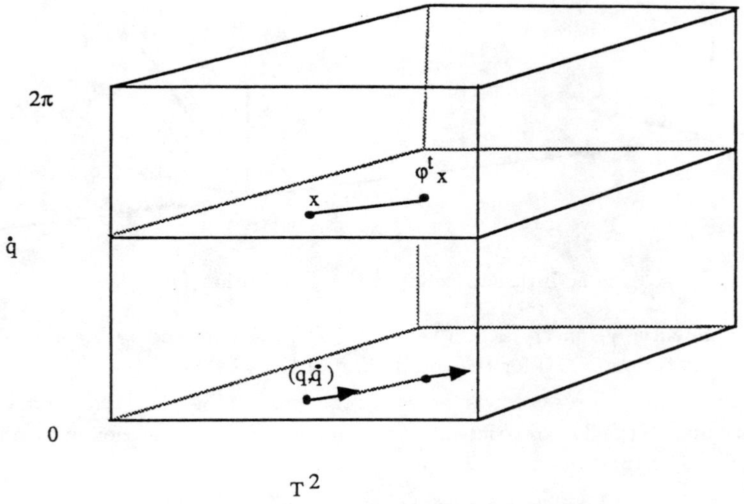

Fig.8. Foliation of M by invariant tori

constant. The phase space, $\mathcal{M} = \{x = (q, \dot{q}) | q \in T^2, |\dot{q}| = 1\}$, which is the torus $\times [0, 2\pi)$, decomposes into a foliation by invariant tori, each torus corresponding to motion in a fixed direction (Fig. 8)

We assign coordinates $\{v, v^\perp, \Psi\}$ to $\mathcal{M}$ in a neighborhood of x: v is the distance from q on the torus, measured in the direction of $\dot{q}$, $v^\perp$ is the distance from q in the direction perpendicular to $\dot{q}$ and Ψ is the angle of the unit velocity vectors measured counterclockwise relative to a fixed axis. These coordinates induce an orthonormal basis $\{X = X_v, X_{v^\perp}, X_\Psi\}$for $T_x\mathcal{M}$. We denote by $P(x)$ the two dimensional subspace of $T_x\mathcal{M}$ spanned by $\{X_{v^\perp}, X_\Psi\}$.

The distribution $P(x)$ gets mapped into itself by the flow:

(16) $$d\psi^t\, P(x) = P(\psi^t x).$$

The evolution of a vector

$$\xi = JX_{v^\perp} + J'X_v \in T_x\mathcal{M}$$

is given by

$$d\psi^t \xi = J(t)X_{v^\perp} + J'(t)X_v \in T_{\psi^t x}\mathcal{M}$$

where

(17) $$J(t) = J_0 + J_0' t \quad \text{and} \quad J'(t) = J_0'.$$

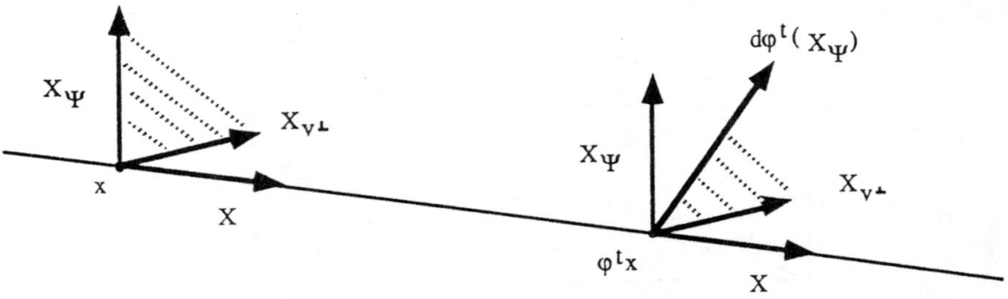

Fig.9. Invariant cones in the distribution $\{P(x)\}$

In differential geometry [Kl], a vector $\xi \in P(x)$ is identified with a Jacobi field, hence the notation (J, J') for the coordinates of the vector.

Equation (17) shows the twist nature of the system (Fig. 9). More precisely, the time one map of the flow, restricted to a two dimensional cross-section perpendicular to the flow is a twist map.

We define the cone $C(x)$ by

$$C(x) = \{J X_{v^\perp} + J' X_\Psi : JJ' \geq 0\}. \tag{18}$$

Under the flow (17), this cone family is invariant but not strictly invariant:

$$\begin{aligned} d\psi^t X_\Psi(x) &\subset \text{Interior}\left(C(\psi^t x)\right) \\ d\psi^t X_{v^\perp}(x) &= X_{v^\perp}(\psi^t x). \end{aligned} \tag{19}$$

The vertical vector has gotten sheared inside the cone, but the horizontal vector stays horizontal.

Interpreting the vectors $\xi \in T\mathcal{M}$ geometrically, we identify a vector ξ with the infinitesimal one-parameter family (variation) of trajectories $\gamma(s) = (v(s), v^\perp(s), \Psi(s))$, $s \in [-\epsilon, \epsilon]$, that generates it: $\gamma(0) = x$ and $\gamma'(0) = \xi$. Our cone is then

$$C(x) = \{\text{variations in } P(x) \text{ that are diverging}\}$$

The edges of the cone are the variations $(J = 1, J' = 0)$ and $(J = 0, J' = 1)$; the former is parallel and is least strongly divergent, the latter is most strongly divergent (Fig. 10). In flat space any diverging variation stays diverging. The parallel family remains parallel though. Thus the cone is invariant but not strictly invariant.

7. Rotation function. What conditions should the rotation function (15) satisfy so that the cone given by (18) will be "kicked up" when it goes through the disk, thereby making the cone field strictly invariant? Geometrically, we want all diverging variations that enter the disk to be strictly diverging when they leave. In particular, we want the parallel family to become strictly diverging. Note that almost every trajectory that starts outside the disk will eventually enter the disk.

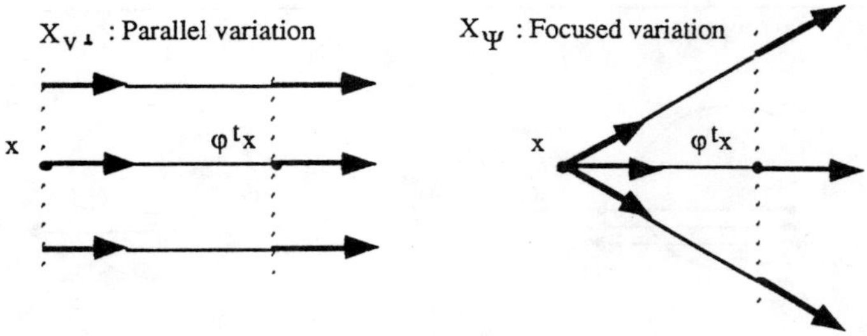

Fig.10. Geometric interpretation of vectors

THEOREM 15 [DL]. *If for almost every $\phi \in (0, \pi)$, the rotation function satisfies either $\Delta\theta'(\phi) > 2$ or $\Delta\theta'(\phi) \leq 0$, then the cone-field given by (18) will become strictly invariant; i.e. any family that is diverging when it enters the disk will be strictly diverging when it leaves. The Lyapunov exponents will be positive almost everywhere and the system will be ergodic.*

Proof. To understand how the cones evolve when they go through the disk, we examine first the case $V(r) \equiv 0$ in the disk. Then the trajectories through the disk would be straight lines. By simple trigonometry, one finds that in this case the rotation function is

$$\Delta\theta(\phi) = 2\phi, \quad \phi \in [0, \pi/2], \tag{20}$$

so that

$$\Delta\theta'(\phi) \equiv 2. \tag{21}$$

In this case, the cones evolve as they had been doing outside the disk. The diverging variations stay diverging, but the parallel variation never becomes strictly divergent; it remains parallel. Thus the cone field does not become strictly invariant.

The trajectories in the parallel variation enter the disk with differing ϕ angles. For $\Delta\theta(\phi) = 2\phi$, the differing amounts these trajectories rotate will just balance out and the trajectories will again be parallel when they leave the disk (Fig. 11a). Knowing the above information in the $V(r) \equiv 0$ case, one can to conclude that if

$$\Delta\theta'(\phi) > 2, \tag{22}$$

then a parallel family that enters the disk with angle ϕ will be strictly divergent when it leaves the disk. The difference in rotation between nearby trajectories will be larger than in the case $\Delta\theta(\phi) \equiv 2\phi$ and hence rather than being parallel when they leave the disk, they will be diverging from one another (Fig. 11b). Also, all

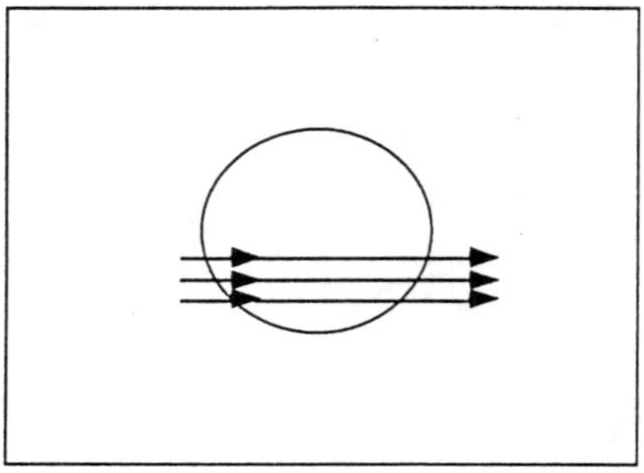

Fig. 11a. $\Delta\theta'(\phi) = 2$

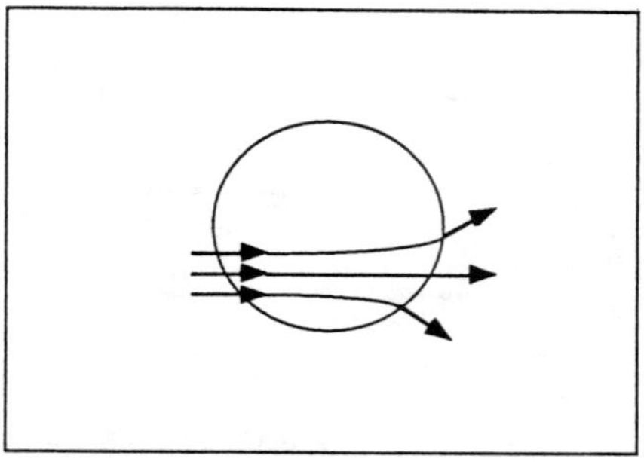

Fig. 11b. $\Delta\theta'(\phi) > 2$

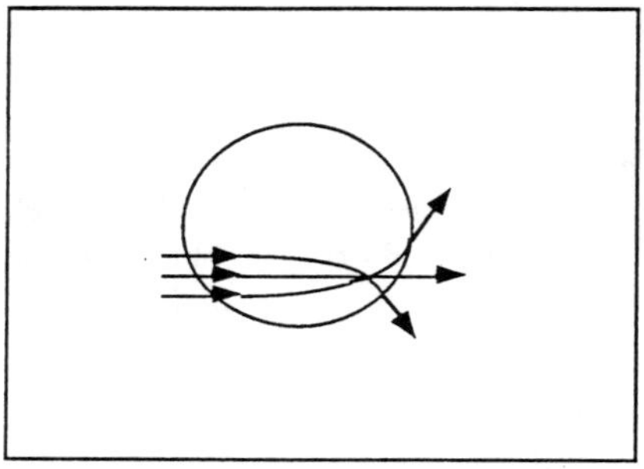

Fig. 11c. $\Delta\theta'(\phi) \leq 0$

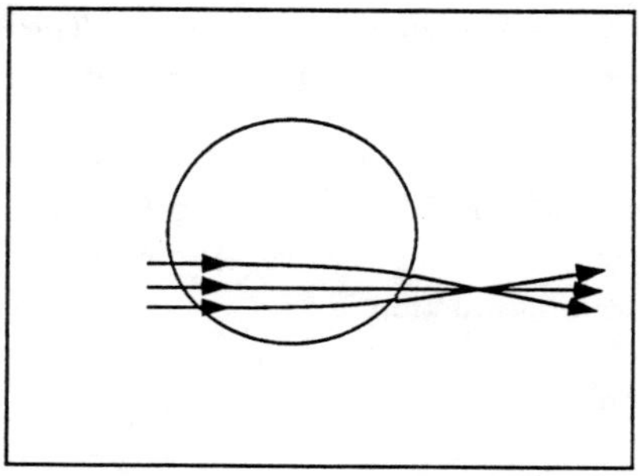

Fig. 11d. $\Delta\theta'(\phi) < 2 - \delta$

the variations that where strictly divergent when they entered the disk will still be strictly divergent when they leave.

If $\Delta\theta'(\phi) \leq 0$, then the same type of geometric analysis shows that any diverging family will be strictly diverging when it leaves the disk. If $\Delta\theta(\phi) > 0$ then the family will have focused while going through the disk (Fig. 11c), which is why it can again be diverging when it leaves the disk.

□

One can perform interesting computer experiments with this model system. If we set the rotation function to be $\Delta\theta(\phi) = 2\phi$, then as we have described above the flow is integrable. If we take a cross-section to the flow and examine the return map to this cross-section with the computer, we find that the phase space is foliated by invariant curves. If we now set the rotation function to be $\Delta\theta(\phi) = c\phi$, $c > 2$, then the above theory implies that the flow will be ergodic. This behaviour is readily seen on the computer: the trajectory of a point goes all over phase space.

Thus it appears that an arbitrarily small perturbation of an integrable system becomes ergodic. One wonders whether this contradicts the K.A.M. theory that asserts that many of the invariant curves should remain under a small perturbation. The resolution of this paradox is that even though the rotation function is changing by an arbitrarily small amount in the C^∞ topology, the underlying potential that produces the rotation function (see §9) is changing in the C^1 topology. Hence K.A.M. theory does not apply.

We can actually show that if

(23) $$\Delta\theta'(\phi) < 2 - \delta$$

for some $\delta > 0$ then the conclusion of Theorem 15 will still hold. However, we must modify our definition of the cones and also the minimum time between returns to the disk must be sufficiently long. The intuitive idea is that under (15), a diverging family might be converging when it leaves the disk. However, given sufficient time the family will focus and then again be diverging when it next returns to the disk. The cone that we define corresponds to variations that are diverging when they enter the disk and are again diverging when they next return to the disk (Fig. 11d).

If the rotation function varies continuously from 2 to values below 2, then we are not able to make an invariant cone-field. The problem is that the diverging variations that enter the disk are converging when they leave the disk, and the time it takes these variations to focus is unbounded. Thus no matter how far apart the disks are placed, there will also be a positive measure set of trajectories for which the diverging variations are not yet diverging by the next return to the disk. Effectively, what has happened is that the expanding direction of the flow has gotten sent to the contracting direction. Present methods can not handle this problem. This problem also occurs in dissipative systems such as the Henon map. In the case of the Henon map, Benedicks and Carlesson [BC] have developed methods which allow them to prove positive Lyapunov exponents.

8. Potential fields on the torus. One way to produce our model systems is to consider a particle of unit mass moving on the torus under the influence of a potential field. The potential in the disk will be rotationally symmetric and vanish on the boundary of the disk. Outside the disk, the potential will be identically zero.

This system serves as a simple model for gas molecules interacting in a box. Sinai [S], extending the work of Krylov [Kr], showed that two hard sphere interacting can be translated into a billiard system consisting of one particle moving on a two dimensional torus T^2 with circular obstacles (scatterers). An elastic collision of the two spheres corresponds to the point particle hitting the scatterer. Here we replace the elastic collision by an interaction governed by a potential.

The motion of this systems is determined by a Hamiltonian function H : $\frac{dp}{dt} = \frac{\partial H}{\partial q}$, $\frac{dq}{dt} = -\frac{\partial H}{\partial p}$, where $p \in R^2$, $q \in T^2$. The Hamiltonian is

(24) $$H(p,q) = \frac{1}{2}p^2 + V(q),$$

where the radially symmetric potential V vanishes outside a disk D of radius R centered at the origin. The total energy E of the point particle is preserved under this flow. We restrict our attention to the $E = 1/2$ energy surface; a compact three dimensional manifold which we denote by $\mathcal{M}$. The choice of $E = 1/2$ implies that the particle will move with unit speed while outside D. We denote by ψ^t the flow induced on $\mathcal{M}$ by (24) and let μ be the restriction of the Liouville measure to $\mathcal{M}$. This measure is invariant under ψ^t.

9. Motion in a rotationally symmetric potential. We assume that any particle that enters the disk will leave the disk. This places certain constraints on the potential. Because the potential is symmetric, the angular momentum of a particle is preserved as it goes through the disk. Hence the system has one integral of motion and is integrable. The symmetry of the potential implies that the motion of a particle between entering and leaving the disk is governed by equation (15). For $\phi \in [0, \pi/2)$, the rotation function is given by

$$(25)\quad \Delta\theta(\phi) = 2\int_{\hat{r}}^{R} \frac{l}{r\sqrt{h(r) - l^2}}\,dr, \quad h(r) = r^2(1 - 2V(r)),\ l = R\cos\phi = h(\hat{r})^{\frac{1}{2}},$$

The lower limit of integration $\hat{r} = r(\bar{t})$ is the minimum distance the particle reaches in its journey through the potential. An orbit that enters the potential field with angle $\pi - \phi$ will rotate clockwise around the disk by the same amount that an orbit entering the disk with angle ϕ will rotate counterclockwise.

Outside the disk, the particle moves in a straight line with unit speed. Thus the dynamics of this system reduces to those of the model system.

Generalizing results of previous authors, we have found several different classes of potentials that satisfy the requirements of Theorem 15 (see [DL] and references therein).

THEOREM 16.

1. For any $\alpha \in (0, 2)$, there exist an attracting potential with singularity of order $-r^{-\alpha}$ for which $\Delta\theta'(\phi) > 2$ for all $\phi \in (0, \pi)$.

2. There exist repelling potentials, which are continuous but not C^1, for which $\Delta\theta'(\phi) < 2 - \delta$ for some $\delta > 0$ and for all $\phi \in (0, \pi)$.

3. There exist smooth potentials without singularities for which $\Delta\theta'(\phi) > 2$ or $\Delta\theta'(\phi) < 0$ for all most every $\phi \in (0, \pi)$.

In the case of elastic reflections (the Sinai billiard), $\Delta\theta(\phi) \equiv 0$ and we have a special case of (2). A natural generalization would be to replace elastic collision with interaction by smooth repelling potentials. For such potentials, the trajectory that just grazes the disk (i.e. "enters" the disk with angle $\phi = 0$) will behave as it does under linear motion. Therefore $\Delta\theta'(0) = 2$, and the values of $\Delta\theta'$ vary continuously from 2 to below 2, leading to a case we can not analysis. By making $V'(R)$ discontinuous, we force the rotation function to satisfy $\Delta\theta'(0) < 2 - \delta$ and then show that $\Delta\theta'(\phi) < \Delta\theta'(0)$.

The potentials in (3) have a critical angle ϕ_{cr} for which $\Delta\theta(\phi_{cr})$ is not defined. For $0 < \phi < \phi_{cr}$, the rotation function satisfies $\Delta\theta'(\phi) > 2$ and for $\phi > \phi_{cr}$, the rotation function satisfies $\Delta\theta'(\phi) < 2$. This discontinuous behaviour in the rotation function is produced by a smooth potential in the following way. Inside the disk, there is a periodic orbit. The trajectory that starts on the boundary of the disk with angle ϕ_{cr} will become asymptotic to this closed orbit and will never leave the disk again.

The results of (3) provide the first example of a smooth Hamiltonian of the form kinetic energy + potential energy for which the flow is ergodic.

10. Geodesic flow. For surfaces with negative Gaussian curvature, the geodesic flow has long been known to be ergodic [H] with positive Lyapunov exponents almost everywhere. This result can only be applied to surfaces of genus $g \geq 2$, since by the Gauss-Bonnet Theorem, $\int_M K(x)\,d\mu(x) = 2\pi(2-2g)$, and hence surfaces of genus $g = 0$ and $g = 1$ can not support metrics of strictly negative curvature.

In [D1] [D2], we showed

THEOREM 17. *There exist C^∞ smooth metrics on the torus and sphere for which the geodesic flow is ergodic with positive Lyapunov exponents almost everywhere.*

We give the torus the flat metric outside the disk. Inside the disk, we make the metric rotationally symmetric. The symmetry of the metric inside the disk implies that the geodesic flow restricted to the disk has an integral of motion, the so called Clairaut Integral. It is the analogue of angular momentum for the potential problem. The motion through the disk in again given by (15). The rotation function is now determined by the metric inside the disk and is given by a formula analogous to (25). The geodesic flow on the two-torus equipped with this metric reduces to our model problem. Thus we must find a metric inside the disk for which the rotation function satisfies either $\Delta\theta'(\phi) > 2$ or $\Delta\theta'(\phi) < 2 - \delta$ for almost every ϕ.

If the metric is smooth, then $\Delta\theta'(0) = 2$ because this is what happens in the flat case and those trajectories that are tangent to the disk (i.e. $\phi = 0$) think they are still on the flat torus. Thus we can not make a smooth metric for which $\Delta\theta'(\phi) < 2 - \delta$ for almost every ϕ. Neither can we make the rotation function satisfy $\Delta\theta'(\phi) > 2$ for almost every ϕ if the rotation function is C^1. For then $\Delta\theta(\pi/2) = \int_o^{\pi/2} \Delta\theta'(\phi)\,d\phi > \pi$, but we know that $\Delta\theta(\pi/2) = \pi$.

The metric we found that produces the desired rotation function can be isometrically embedded in R^3 (Fig. 12). It first has negative curvature (a collar) and then a cap (focusing cap), which in its simplest form, has positive curvature. Separating the two regions, at the base of the cap, is closed geodesic lying in zero curvature. Geodesics that enter the disk with a certain critical angle ϕ_{cr} will become asymptotic to the closed geodesic at the base of the cap and will never leave the disk. For $\phi < \phi_{cr}$, the rotation function will satisfy $\Delta\theta'(\phi) > 2$. Geometrically, the corresponding geodesics will stay in the negative curvature collar, never rising above the base of the cap. When going through negative curvature, every diverging variation

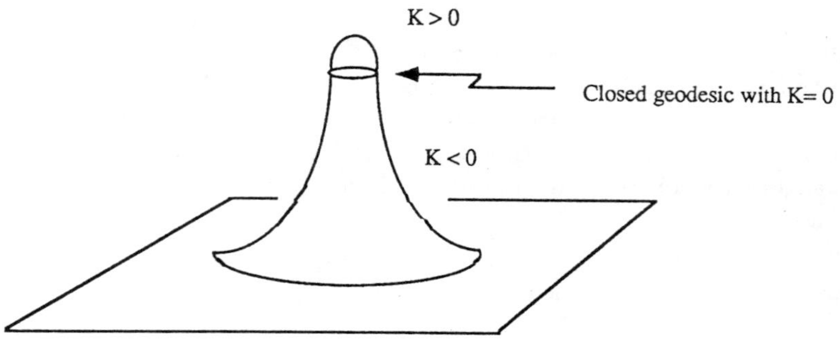

Fig.12. Flat torus with cap

stays diverging and, most importantly, the parallel family becomes strictly divergent. For $\phi > \phi_{cr}$, the rotation function satisfies $\Delta\theta'(\phi) < 0$. The corresponding geodesics rise through the collar and enter the cap. In going through the negative curvature in the collar, all the diverging variations become strictly divergent. The cap is constructed in such a way that any strictly divergent variation that enters the cap will focus once while in the cap then be strictly divergent when it leaves the cap. Descending the collar, these strictly diverging variations stay strictly divergent . Hence when they leave the disk and return to the flat part of the torus, they are strictly divergent. This behaviour is analogous to case (3) for potentials. We call this surface "flat torus with cap".

For the sphere, the Gauss-Bonnet theorem forces us to proceed in a slightly different manner. We delete three (or more) points from the sphere. The universal cover of this surface is the disk and we can project the Poincare metric of curvature $K \equiv -1$ onto the surface. In a neighborhood of the deleted points this metric is embeddable, giving a rotationally symmetric cusp that goes to infinity. To complete the metric on the sphere, we cut off the end of the cusp and attach the focusing cap. We can perform this attachment in a C^∞ way and so that in the transitional region that connects the cap to the $K \equiv -1$ region the curvature is negative.

We define the cone of diverging variations for points that start outside the cap by (18). The basis vectors $\{X_{v^\perp}, X_\Psi\}$ again correspond to the infinitesimal families that are parallel and that are focused. The latter is now determined using covariant differentiation and parallel transport rather than a local coordinate system. If we give the base of the cap coordinates $(\tilde{\theta}, \tilde{\phi})$ then the net effect of going through the cap is

$$(\tilde{\theta}, \tilde{\phi}) \to (\tilde{\theta} + \tilde{\Delta}\theta(\tilde{\phi}), -\tilde{\phi}), \tag{26}$$

where $\tilde{\Delta}\theta$ is the rotation function from entering to leaving the cap. The special focusing cap we have constructed has the property that $\tilde{\Delta}\theta'(\tilde{\phi}) < 0$. Hence going through the cap is another integrable twist map.

We think of this geodesic flow as being a composition of a hyperbolic system (the geodesic flow in the $K < 0$ region) and an integrable twist system (the geodesic flow in the rotationally symmetric cap). For this system, it is easier to make the cone-fields strictly invariant than it is for the case of flat torus with cap. In negative curvature, the parallel variation will immediately become strictly divergent. Thus the cones (18) will go strictly inside themselves when they are outside the cap. Once the cones have become strictly invariant outside the cap, passing through the cap serves to preserve the cones.

The author first made his construction for the sphere. At the time, his idea was that the negative curvature would produce expansion (cones going inside themselves) and that going through the cap would just preserve the directions of expansion (diverging variation goes to diverging variation). To prove that all the diverging variations got made to focus in the cap, he needed to have very explicit control over how the geodesics in the cap behaved. That is why he used a surface of revolution (an integrable system). Only later did he look at the problem from the reverse perspective : going through the cap (equation (26)) preserved the cone-fields and only a small nudge was needed to make the parallel family become strictly diverging; i.e. a twist transverse to the original twist. This nudge was achieved by the collar of negative curvature. Once diverging variations had become strictly divergent, then this property would be preserved by zero curvature (the flat torus).

11. A new metric on the torus with ergodic geodesic flow. In the linked twist maps described in [BE][W1], the component twist maps were integrable. This condition is not necessary for the linked twist map to have positive Lyapunov exponents almost everywhere. For example in the billiard systems described in §4, the map from entering to leaving a focusing arc is a non-integrable twist map.

If the first twist map is integrable, then in appropriate coordinates the horizontal edge of the cone stays horizontal. If the twist of the second map is suitably aligned, it will "kick up" this edge of the cone so that it goes strictly inside the image cone. If the horizontal edge of the cone were to drop a small amount, rather than staying horizontal, then the second twist map could still serve to "kick up" the edge enough to make it go inside the image cone. This trick ceases to work when the horizontal edge drops too much.

Using this reasoning, we generalize the flat torus with cap construction to produce a metric on the two-torus that has variable curvature outside a union of disks and for which Theorem 17 holds.

We start by placing a finite number of disks, $D_i, i = 1, \ldots, n$, on the flat torus in such a way that every geodesic will enter (i.e. $\phi > 0$) one of disks in a bounded time. This restriction is called the finite horizon condition (see for example [B4]). Inside each disk, we put the same metric that was used in the flat torus with cap construction. Remember that this metric satisfies $\Delta\theta'(\phi) < 0$ or $\Delta\theta'(\phi) > 2$ for all $\phi \in \{(0, \pi) \setminus \phi_{cr}\}$ and $\Delta\theta'(0) = 2$.

We place the disks in such a way that no trajectory can be tangent to more than two disks in succession. Then by compactness, there is some finite time τ and some number δ such that every trajectory that starts outside the disks will enter a disk with angle $\phi \in (\delta, \pi - \delta)$ in time $t < \tau$. Here we measure the time only when the trajectory is outside the disks. The cones given by (18) will become strictly invariant in time $t < \tau$ and most importantly the edges of the cones will go inside by a uniform amount. That is, there exist constants $0 < c_1 < c_2 < \infty$ such that for $x \in \mathcal{M} \setminus \bigcup_i D_i$ and $\xi \in \mathcal{C}(x)$,

$$c_1 < m\,(d\psi_x^\tau \xi) < c_2. \tag{27}$$

We now perturb the metric outside $\bigcup_i D_i$. By continuity, any sufficiently small (depending on τ, c_1), smooth perturbation will produce a geodesic flow $\tilde{\psi}^t$ for which $0 < m\left(d\tilde{\psi}_x^\tau \xi\right) < 2c_2$. Thus we can apply Theorem 1 and 2 to this system and conclude that Theorem 17 still holds. The resulting metric will have a mixture of both positive and negative curvature outside $\bigcup_i D_i$.

Acknowledgements. The author benefited from helpful discussions with Mike Jolly, who was responsible for the computing described in §7, Andreas Knauf and Carlangelo Liverani. He thanks the IMA for its friendly hospitality during his stay in spring, 1990.

REFERENCES

[B1] L.A. Bunimovich, *On the ergodic properties of nowhere dispersing billiards*, Commun. Math. Phys. 65 (1979), 295-312.

[B2] L.A. Bunimovich, *A theorem on ergodicity of two-dimensional hyperbolic billiards*, Commun. Math. Phys. 130 (1990), 599-621.

[B3] L. A. Bunimovich, *On absolutely focusing mirrors, preprint.*

[B4] L. A. Bunimovich, *Decay of correlations in dynamical systems with chaotic behavior*, Sov. Phys. JETP 62(4), October 1985.

[BC] M. Benedicks and L. Carleson, *The dynamics of the Henon map*, preprint.

[BE] R. Burton and R.W. Easton, *Ergodicity of linked twist maps*, Springer Lecture Notes in Math., Vol. 819 (1980), 35-49.

[BG] K. Burns and M. Gerber, *Continuous invariant cone families and ergodicity of flows in dimension three*, Ergod. Th. & Dynam. Sys. 9 (1989), 19-25.

[BS] L.A. Bunimovich and Ya. G. Sinai, *Statistical properties of Lorentz gas with periodic configuration of scatterers*, Commun. Math. Phys. 78 (1981) 479-497.

[CFS] I.P. Cornfeld, S.V. Fomin, Ya. G. Sinai, *Ergodic Theory*, Springer Verlag, 1982.

[D1] V.J. Donnay, *Geodesic flow on the two-sphere, Part I: Positive measure entropy*, Ergod. Th. & Dynam. Sys. (1988), 8, 531-553.

[D2] ——————, *Geodesic flow on the two-sphere, Part II: Ergodicity*, Dynamical Systems, Springer Lecture Notes in Math., Vol. 1342 (1988), 112-153.

[D3] ——————, *Using integrability to produce chaos: billiards with positive entropy*, Commun. Math. Phys (to appear).

[DL] V. J. Donnay and C. Liverani, *Potentials on the two-torus for which the Hamiltonian flow is ergodic*, Commun. Math. Phys. 135 (1991), 267-302.

[H] E. Hopf, *Statistik der geodatischen Linien in Mannigfaltigkeiten negativer Krummung*, Leipziger Berichte 91 (1939), pp. 261-304.

[KS] A. Katok and J.-M. Strelcyn, *with the collaboration of F. Ledrappier and F. Przytycki, Invariant manifolds, entropy and billiards; smooth maps with singularities*, Springer Lecture Notes in Math., Vol. 1222 (1986).

[Kl] W.Klingenberg, *Riemannian Geometry*, de Gruyter Studies in Mathematics, 1982.

[Kr] N.S. KRYLOV, *Works on the foundations of statistical physics*, Princeton University Press, 1979.

[KSS] A. KRAMIL, N. SIMANYI, D. SZASZ, *A transversal theorem for semi-dispersing billiards*, Commun. Math. Phys., 129 (1990), pp. 535–560.

[L] V.F. LAZUTKIN, *On the existence of caustics for the billiard ball problem in a convex domain*, Math. USSR Izv. 7 (1973), pp. 185–215.

[Mr1] R. MARKARIAN, *Billiards with Pesin region of measure one*, Commun. Math. Phys 118 (1988), 87-97.

[Mr2] R. MARKARIAN, *Non-uniform hyperbolicity, quadratic forms and billiards*, preprint.

[O] V.I. OSELEDEC, *The multiplicative ergodic theorem. The Lyapunov characteristic numbers of dynamical systems*, Trans. Mosc. Math. Soc. 19 (1968), 197-231.

[P] YA.B. PESIN, *Lyapunov characteristic exponents and smooth ergodic theory*, Russ. Math. Surveys 32(1977), 55-114.

[S] YA.G. SINAI, *Dynamical systems with elastic reflections. Ergodic properties of dispersing billiards*, Russ. Math. Surv. 25 (1970), 137-189.

[SC] YA. G. SINAI AND N.I. CHERNOV, *Ergodic properties of certain systems of two-dimensional disks and three-dimensional balls*, Usp. Math. Nauk 42 (1987), 153-174.

[W1] M. WOJTKOWSKI, *Linked twist mappings have the K-property. Annals of the New York Academy of Sciences Vol. 357*, Nonlinear Dynamics (1980), 65-76.

[W2] M. WOJTKOWSKI, *Invariant families of cones and Lyapunov exponents*, Ergod. Th. & Dynam. Sys. 5 (1985), 145-161.

[W3] M. WOJTKOWSKI, *Principles for the design of billiards with nonvanishing Lyapunov exponent*, Commun. Math. Phys. 105 (1986), 319-414..

GHOST TORI FOR MONOTONE MAPS

CHRISTOPHE GOLÉ*

Abstract. Certain higher dimensional analogs of twist maps on $\mathbb{T}^n \times \mathbb{R}^n$ come equipped with a discrete variational problem. We study the gradient flow of the action and find invariant sets with the cohomology of $\mathbb{T}^n$. This implies a topological lower bound on the number of periodic orbits of each given prime rotation vector. For the twist map case, one can extract circles that interpolate Aubry-Mather sets from these invariant sets.

Key words. Symplectic, monotone maps, ghost tori, Aubry-Mather sets

0. Introduction. This article is a survey of the author's work on twist maps of $\mathbb{S}^1 \times \mathbb{R}$ and their higher dimensional symplectic analog on $\mathbb{T}^n \times \mathbb{R}^n$, called monotone maps ([G 1,2,3]). With such a map F is associated a "discrete lagrangian" on the space $(\mathbb{R}^n)^{\mathbb{Z}}$ of bi-infinite sequences whose critical points correspond one to one to the orbits of F. This is analogous to the Hamiltonian setting where one finds periodic orbits as critical points of the action on the loops space of the manifold. Our main result is a variation of the Arnold's conjecture [Ar] which states that certain symplectic maps on a manifold have at least as many fixed points as Morse theory garanties critical points for a real valued function on this manifold.

More precisely, we find at least $cl(\mathbb{T}^n) = n + 1$ periodic orbits of any given prime rotation vector and $sb(\mathbb{T}^n) = 2^n$ if they are nondegenerate. Since $\mathbb{T}^n \times \mathbb{R}^n$ is not compact, we need some boundary condition at infinity on our maps, e.g. that they be completely integrable there, or not too far from it. Other than that, the maps can be as far as we want from completely integrable. In fact, the result is valid for arbitrary finite compositions of these maps. This is one of the differences between our work and the one of Bernstein and Katok [B-K], whose setting we otherwise very much adopted. Another marked difference with their work is that we drop their convexity assumption, replacing it by a more general nondegeneracy condition. Herman [He] has shown that the dynamics of monotone maps can differ drastically whether one assumes or not this convexity, especially regarding invariant tori (our result covers what he calls the indefinite case).

To a certain extend, we exploit the analogy between this setting and the Hamiltonian one to obtain our result. We build isolating blocks with the same properties as those appearing in the seminal work of Conley and Zehnder [C-Z 1]. One problem arises when we want to count periodic *orbits* and not only points. One has to take a quotient of the isolating block by the group $\mathbb{Z}_q$ where q is the period. In some cases, some topological information is lost in the process.

To remedy this situation, we also consider monotone maps that continue to completely integrable ones through a curve of monotone maps. This enables us to use Floer's theorem of continuation of normally hyperbolic invariant set that he used in his thesis [Fl 1], [Fl-Z] to prove Arnold's conjecture for surfaces. This

*Department of Mathematics, State University of New York at Stony Brook, Stony Brook, NY 11794-3651

theorem implies in our case that the invariant torus of rotation vector p/q, *seen as completely critical invariant set for the energy flow* in the space of periodic sequences of rotation vector p/q, can be continued in this space as an invariant set for the flow, in the sense of Conley [Co] and that it conserves the cohomology of $\mathbf{T}^n$. For this reason, we call such an invariant set a *ghost torus.*

It is not known to this author whether or not all monotone maps can be continued to a completely integrable one through a time periodic Hamiltonian flow. If this were the case, our result would derive from the analogous one of Josellis [J]. However, we hope to show that the techniques exposed here can be of some use other than for the existence of periodic orbits.

One question which remains open is whether or not one can find orbits of all rotation vectors. Note that this is weaker than asking for quasiperiodic orbits, on which the action of the map would be conjugated to a rigid rotation. In our setting, we conjecture the existence of a ghost torus for all given rotation vector. We think that this is the first step in order to answer positively to the above question. To support this conjecture, we prove it in the case of twist maps. For this, we use the monotonicity of the energy flow associated to twist maps ([An]) to develop the concepts of *ghost circles* and more generally of *sigma-Aubry-Mather sets* (abr. σAM sets). The σAM sets are subsets of $\mathbf{R}^{\mathbf{Z}}$ that are closed, completely ordered, shift and integer translation invariant (The partial order on $\mathbf{R}^{\mathbf{Z}}$ is the usual one , given by the positive cone). In a sense, they are all the potential Aubry-Mather sets for twist maps. They are homeomorphic to lifts of closed invariant sets of circle homeomorphisms. The rotation number is well defined and is continuous on the set of σAM sets. Aubry-Mather sets can be defined as critical σAM sets and ghost circles as connected, flow invariant σAM sets.

Our second result shows that any Aubry-Mather set (as defined above) can be embedded in a ghost circle. A corollary of this is the existence of ghost circles of all given rotation number, i.e. a positive answer to our conjecture.

Finally we indicate how ghost circles might useful to understand the ΔW of Mather and how the properties of ghost circles could be used in the theory of transport.

The author would like to thank warmly the numerous participants of the dynamical systems year at the IMA who helped him complete this work.

1. Notation and statement of the main result. We let $\mathbf{T}^n$ be $\mathbf{R}^n/\mathbf{Z}^n$ and consider the space $\mathbf{A}^n := \tilde{\mathbf{T}}^n \times \mathbf{R}^n \cong \mathbf{R}^n \times \mathbf{R}^n$ with (global) coordinates (x,y). The group $\mathbf{Z}^n$ acts as deck transformations by

$$T_m(x, y) = (x + m, y), \quad m \in \mathbf{Z}^n.$$

$\mathbf{A}^n$ is endowed with the canonical symplectic form

$$\Omega = \sum_{k=1}^{n} dx_k \wedge dy_k = d\alpha, \text{ where } \alpha = \sum_{k=1}^{n} y_k dx_k.$$

A diffeomorphism F of A^n is *symplectic* if

$$F^*\Omega = \Omega,$$

and *exact symplectic* if $F^*\alpha - \alpha$ is exact, that is if:

$$F^*\alpha - \alpha = dS,$$

where S is a C^2 real valued function on A^n. Obviously, an exact symplectic map is symplectic.

In the following, we write $F(x,y) = (X,Y)$.

DEFINITION 1 (MONOTONE MAPS, GENERATING FUNCTIONS).

A diffeomorphism F of A^n is called a *monotone map* if:

(1) F is exact symplectic: $F^*\alpha - \alpha = dS, \quad S : A^n \to \mathbb{R}$
(2) F is a lift: $F \circ T_m = T_m \circ F$
(3) For each x_0,the map $(x_0, y) \to (x_0, X)$ is a diffeomorphism of $\mathbb{R}^n$ (and hence $(x,y) \to (x,X)$ is a diffeomorphism of $\mathbb{R}^{2n}$)

The function S above is called the *generating function* of the monotone map, thought of as a function of (x, X).

That $(x,X) \to (x,y)$ is a diffeomorphism implies the nondegeneracy condition:

$$det \partial_1 \partial_2 S(x,X) \neq 0$$

which generalizes the so-called twist condition (∂_1 (resp. ∂_2) means the partial derivative with respect to the first (resp. the second) component). We will see in the next section in what sense S generates F.

When $n = 1$, we have:

DEFINITION 2 (TWIST MAPS).

A diffeomorphism f of A is called an area preserving monotone twist map, or, in short (in this paper), twist map if it is monotone and if the scalar $-\partial_1\partial_2 S(x,X)$ is strictly positive for all (x, X).

COMMENTS.

(1) Definition 1.3 is slightly different from the one that we used previously [Go 1,2] where we put condition 1.4 map $(x,X) \to (x,y)$ is a diffeomorphism. The term monotone was introduced by Herman ([He]) in a more general setting : his monotone maps do not necessarily have a generating function. Some authors impose another condition on S, i.e. that $-\partial_1\partial_2 S$ be positive definite. They call these maps symplectic twist maps (see e.g.[K-M]). Definition 1.1, on the other hand, is equivalent to the usual one: f is an area preserving map with zero flux which sends any vertical line

$x = c$ into a graph over the x-axis satisfying $f(c, +\infty) = (+\infty, +\infty)$ and $f(c, -\infty) = (-\infty, -\infty)$.

(2) Note that some authors use "exact symplectic" for maps that are homologous to the identity, i.e., time one maps of (time dependant) Hamiltonian flows (what E.Zehnder now calls Hamiltonian maps). Whereas for the case $n = 1$ (twist maps) it is true that the maps we consider can be suspended by a hamiltonian flow ([Mo1]), it seems to be an open question whether this holds for the monotone maps we defined above.

The type of objects that we are looking for are:

DEFINITION 3 ((P,Q)-PERIODIC POINTS). Let (p,q) be an element of $\mathbb{Z}^n \times \mathbb{Z}$. (x_0, y_0) in $\mathbb{A}^n$ is a (p,q)-periodic point ((p,q)-point, in short) if, when we denote by $(x_k, y_k) = F^k(x_0, y_0)$, the following holds:

$$x_{k+q} = x_k + p$$

The orbit of a (p,q)-point is called a (p,q)-*orbit.* (p,q) is called relatively prime if q is prime with at least one of the components of p.

As stated in the introduction, in order to find such orbits, we need to assume a boundary condition for our map. It is better expressed in terms of the generating function S.

BOUNDARY CONDITIONS. We suppose that S can be written:

$$(\partial) \qquad \begin{aligned} S(x, X) &= S_0(x, X) + R(x, X), \text{ with} \\ S_0(x, X) &= < A(X - x), (X - x) > \end{aligned}$$

where A is a symmetric nondegenerate matrix and R assumes sublinear growth in its derivative:

$$\lim_{\|X-x\| \to \infty} \frac{\|\partial_\alpha R_\lambda(x, X)\|}{\|X - x\|} = 0, \ \alpha = 1, 2$$

As stated in the introduction, this boundary condition implies that the map is somewhat comparable to a completely integrable one (generated by S_0) at infinity. However, this boundary condition is flexible enough to allow $R(x, X) = \nabla V(x)$ for any periodic V , and it is enough that this be satisfied for $X - x$ large. This includes, in particular, maps which are in the standard family (see the examples below). In particular, this includes twist maps without any invariant (homotopically non trivial) circles . *Note that A is not assumed to be positive definite.* When it is not, F falls into the class of indefinite maps considered by Herman [He].

CONTINUATION SETTING. *We say that a monotone map F with generating function S continues to a completely integrable F_0 generated by S_0 if there is a family*

$$S_\lambda(x, X) = S_0(x, X) + R_\lambda(x, X)$$

generating monotone maps with $R_0 = 0$ and $S_1 = S$.

We can now state our main result.

THEOREM 1. *Let F be a monotone map continuing to a completely integrable one and satisfying the boundary condition (∂). Then F has at least $n+1$ distinct (p,q)-periodic orbit for each prime (p,q) in $\mathbb{Z}^n \times \mathbb{Z}$, and 2^n of them if they are nondegenerate.*

Remember that a q-periodic orbit is nondegenerate if $det(DF^q(Z) - Id) \neq 0$ for any z in the orbit. The above theorem remains true for compositions of F_i's satisfying the conditions of the theorem, as long as their matrices A_i commute and are of the same sign on their common eigenspaces.

If we do not assume the continuation setting, we have the following:

THEOREM $1'$.

Let $F = F_1 \circ \ldots \circ F_s$ where each F_i is a monotone map of $\mathbb{A}^n$ generated by a global generating function S_i. Suppose that each F_i satisfies the boundary condition ∂ and that the corresponding set $\{A_i\}_{i=1}^s$ of symmetric, nondegenerate matrices is such that

$$[A_i, A_j] = 0 \quad \forall i, j \in \{1, \ldots, s\},$$

and such that all the A_i's are positive definite (resp. negative definite) in the same subspaces of dimension k_0 (resp. $n - k_0$). Then:

(1) *F has at least $n+1$ geometrically distinct (p,q)-periodic orbits for each prime (p,q) in $\mathbb{Z}^n \times \mathbb{Z}$.*

Suppose in the following that all (p,q)-periodic orbits are nondegenerate. Then:

(2) *If either q is odd, s is even, or k_0 is even or equal to n or 0, then F has at least 2^n (p,q)-periodic orbits. Otherwise F has at least 2^{n-1} of them.*

Note that theorem 1' is stronger than theorem 1 only in the sense that it does not assume the continuation setting. Otherwise, we see that some information is lost when we are seeking nondegenerate orbits in this more general setting. The proof of $1'$ is noticeably more complicated than the one of 1, which has also the advantage to yield to the concept of ghost torus. It would be interesting to find out if all monotone maps continue to a completely integrable one.

RELATED RESULTS. The above theorems are in direct line with the Poincaré Birkhoff theorem for twist maps, see [G-H] for the historical background and recent generalizations. The Birkhoff-Lewis theorem finds infinitely many periodic orbits in the neighborhood of an elliptic fixed point of a symplectic map. Transcribed into polar coordinates, it is a theorem of existence of periodic orbits for the perturbation of a completely integrable map. Moser [Mo 2] gave a complete and elegant proof of this theorem. The breakthrough for global results in higher dimensions came with the theorem of Conley and Zehnder [C-Z 1]. They consider the setting of time-one maps of time periodic Hamiltonian systems on $\mathbb{T}^n \times \mathbb{R}^n$ (as well as on $\mathbb{T}^{2n}$) and assume that they are completely integrable outside a bounded set. They find $n+1$ (or 2^n nondegenerate) ***homotopically*** trivial periodic orbits of period 1. Josellis [J] refined this result by proving a theorem very much like ours, in the Hamiltonian setting.

Bernstein and Katok [B-K] used the discrete variational approach to find $n+1$ periodic orbits of any given type for the perturbation of a "convex" monotone map. They came close to proving a result of regularity on the orbits they find which would have enabled them to find points with irrational rotation vector. Chen studied the indefinite case and found $n+1$ periodic *points* in an otherwise similar setting to that of [B-K].

EXAMPLES.

(1) Completely integrable and standard monotone maps:

$$F_0(x,y) = (x+y,y)$$

has the generating function $S_0(x,x') = \frac{1}{2}(x'-x)$ and its energy flow (see next section) is given by the infinite system of O.D.E.'s:

$$\dot{x}_k = -2x_k + x_{k-1} + x_{k+1},$$

i.e. a discretised heat equation.

One can add a "potential" to S_0 : let

$$S(x,x') = \frac{1}{2}(x,x') - g(x).$$

Then S generates $F(x,y) = (x+y+h(x), y+h(x))$ where $h(x) = g'(x)$. The associated flow is given by:

$$\dot{x}_k = -2x_k + x_{k-1} + x_{k+1} + h(x_k).$$

Angenent remarked that this was the discretisation of $x_s = x_{tt} + h(t,x)$ and proved important results for twist maps by developping this analogy. When $n=1$ and

$$g(x) = -\frac{k}{4\pi^2} cos(2\pi x) \quad ,$$

F is called the *standard map* (or standard family).

(2) Hamiltonian systems:

Let $H(x,y,t) : \mathbb{R}^n \times \mathbb{R}^n \times \mathbb{R} \to \mathbb{R}$ be periodic in x and t. If we assume

$$Det\frac{\partial^2 H}{\partial y^2} \neq 0,, \text{ and that, for } |y| > a, \quad H(x,y) = \frac{1}{2} < Ay, y > + < C, y >$$

for A symmetric, nondegenerate, and $C \in \mathbb{R}^n$ (see Conley-Zehnder, Thm 3 [C-Z 2]) then the time-1 map F can always be decomposed into monotone F_i's. Moreover, they trivially satisfy the boundary conditions of our theorem. Also note that such Hamiltonian systems fall into the domain of those studied by Josellis [J]

2. The variational setting. The function S generates a monotone F in the following (classical mechanic) sense:

$$\begin{aligned} y &= -\partial_1 S(x, X) \\ Y &= \partial_2 S(x, X) \end{aligned}$$

Let $z_k = F^k(z_0) = (x_k, y_k)$. The orbit $\{z_k\}$ is completely determined by the sequence x_k of $(\mathbb{R})^{\mathbb{Z}}$. Indeed, we deduce:

$$y_k = -\partial_1 S(x_k, x_{k+1}) = \partial_2 S(x_{k-1}, x_k)$$

This can be written:

$$\partial_1 S(x_k, x_{k+1}) + \partial_2 S(x_{k-1}, x_k) = 0$$

which can be formally interpreted as:

$$(*) \qquad \begin{aligned} &\nabla W(\mathbf{x}) = 0, \quad \text{for} \\ &W(\mathbf{x}) = \sum_{-\infty}^{+\infty} S(x_k, x_{k+1}) \text{ and } \mathbf{x} \in (\mathbb{R}^n)^{\mathbb{Z}}. \end{aligned}$$

What we have proven here is that there is a 1-1 correspondance between points of $\mathbb{A}^n$ and the critical points of ∇W in $\mathbb{R}^{\mathbb{Z}}$. In particular, invariant tori for the map correspond to completely critical tori for the vector field ∇W.

One can think of the above construction as a discrete version of the classical one: the map $(x, X) \to (x, X)$ is the analog to the Legendre transformation ($X - x$ is the discretised velocity) and equation $(*)$ is a formulation of the "least action principle".

Of course, W is not well defined, since the sum is not convergent. However, "∇W" is well defined and generates a flow on $(\mathbb{R}^n)^{\mathbb{Z}}$ that we call the energy flow.

More precisely when S is a C^p function, the infinite system of O.D.E's:

$$-(\nabla W(\mathbf{x}))_k = \dot{x}_k = -[\partial_1 S(x_k, x_{k+1}) + \partial_2 S(x_{k-1}, x_k)]$$

defines a C^{p-1} local flow ζ^t on $(\mathbb{R}^n)^Z$ (with the usual product topology) whose critical points are in one to one correspondance with the points of $\mathbb{A}$ and their orbit under the map. This flow can also be made global by assuming relevant boundary conditions (e.g. that the map be completely integrable outside a bounded strip, or at infinity).

Because F is a lift, S is periodic: $S(x + m, X + m) = S(x, X)$, for any $m \in \mathbb{Z}^n$ and hence the flow ζ^t can be defined on $(\mathbb{R}^n)^Z/Z$

3. Sketch of the proof and ghost tori. To look for (p,q)-periodic points, one restricts the flow ζ^t to the set of periodic sequences :

$$X_{p,q}(\mathbb{R}^n) = X_{p,q} := \{\mathbf{x} \in (\mathbb{R}^n)^{\mathbb{Z}} \quad | \quad x_{k+q} = x_k + p\}$$
$$\cong \{(x_0, \dots, x_{q-1}) \in (\mathbb{R}^n)^q\}$$

Because of the periodicity of S, it is easy to see that $X_{p,q}$ is invariant under ζ^t. Moreover ζ^t is in fact the gradiant flow of the well defined function:

$$W_{p,q}(\mathbf{x}) = \sum_{k=0}^{q-1} S(x_k, x_{k+1})$$

with respect to the coordinates $(x_0, \dots, x_{q-1})$ of $X_{p,q}$ (with the convention that $x_q = x_0 + p$).

Hence critical points of ζ^t in $X_{p,q}$ correspond 1-1 to (p,q)-points.

In our thesis, we prove that, given the boundary condition ∂, the following set is an isolating block for ζ^t, whenever C is large enough:

$$B(C) = \{(\mathbf{x}) \in X_{p,q} | \sup_{k \in \{1, \dots, n\}} \sum_{i=1}^{q} (x_i^k - x_{i-1}^k)^2 \leq C^2\}$$

It is not hard to see that $B(C)$ is homeomorphic to the product of $\mathbb{T}^n$ with the product of closed disks. To show that it is an isolating block, one has to show that the flow enters some of these disks and exit others (in the positive definite case it always enters, i.e. we have an attracting block). This situation is quite similar to the one encountered in the original proof of Conley and Zehnder [C-Z 1], where they use their generalised Morse theory to find $n+1$ critical points or 2^n if nondegenerate for a gradiant flow with a similar isolating block.

However, a problem arises when one wants to count the distinct periodic *orbits* instead of points. One has to take a quotient of $X_{p,q}$ by the shift σ on sequences ($\{\sigma\mathbf{x}\}_i = x_{i+1}$ which acts as $\mathbb{Z}_q$ on $X_{p,q}$). In our thesis, we proved a theorem of equivariance of the Conley index to study the topology of the quotient $\underline{B}(C)$ of $B(C)$. Information is lost when counting nondegenerate points with the Conley-Zehnder Morse inequalities whenever the quotient map is not a trivial fibration. In these cases, we have to take a double cover, which explains the 2^{n-1} in theorem 1′.

In the continuation setting, we take another route: instead of relying on the topology of the index pair $(B(C), B^-(C))$ (where $B^-(C)$ is the exit set), we use the idea of Floer ([Fl], see also [F-Z]) in his thesis to prove the following a topological continuation theorem for the invariant set contained inside $B(C)$.

DEFINITION 4. We denote by G_{pq}^{λ} and call *ghost torus* the maximum invariant set in $B(C)$ for the energy flow induced by the generating function S_λ.

THEOREM 2 (CONTINUATION OF THE GHOST TORI).

(1) *G_{pq}^0 is the critical torus corresponding to the F_0-invariant torus in $\mathbb{A}^n$ with rotation vector p/q.*
(2) *G_{pq}^0 is normally hyperbolic for the flow induced by S_0.*

(3) *There is a retraction* $r : X_{p,q} \to G^0_{pq}$.

(4) *Since* C *can be chosen so that* $B(C)$ *is an isolating block for the energy flows induced by each of the* S_λ*'s, the ghost tori* G^λ_{pq} *are related by continuation in the sense of Conley [Co].*

(5) *In conclusion, Floer's result implies:*

$$(r\mid_{G^\lambda_{pq}})^* : H^*(G^0_{pq}) \to H^*(G^\lambda_{pq})$$

is an injective map from the cohomology of $\mathbf{T}^n$ *into that of* G^λ_{pq}.

(6) *All of the above remain true when taking the quotient by the action of* σ.

In other words, the topology of G^λ_{pq} can only get more complicated than the one of $\mathbf{T}^n$ as the parameter varies. For this reason we will in general refer to a compact invariant set with this property as a ghost torus.

The important advantage of this approach is that *all the above feature are conserved when we quotient by the action of* σ. Hence we can look at the ghost tori $\underline{G}^\lambda_{pq}$ in the quotient space: their topology remain always as complicated as that of $\mathbf{T}^n$. As a corollary, using Conley-Zehnder Morse theory [C-Z 1,2] on the invariant set $\underline{G}^1_{pq}$, we get at least $cl(\mathbf{T}^n) = n+1$ critical points and $sb(\mathbf{T}^n) = 2^n$ if they are nondegenerate. *These critical points correspond to distinct orbits of* F. In the next sections, we will see how the ghost tori may be of interest in themselves.

4. Irrational rotation vectors?

One way to attack the problem of finding orbits of all rotation vectors would be to consider the flow $\dot{\mathbf{x}} = -\nabla W(\mathbf{x}) = \partial_1 S(\mathbf{x}_i, \mathbf{x}_{i+1}) + \partial_2 S(\mathbf{x}_{i-1}, \mathbf{x}_i)$,

$$\mathbf{Y} = \bigcup_{\omega\in\mathbf{R}^n} Y_\omega \text{ where}$$

$$Y_\omega = \{(\mathbf{x}) \in (\mathbf{R}^n)^{\mathbf{Z}} \quad \mid |\mathbf{x}_j - j\omega| < \infty\}$$

One can check that Y_ω is invariant under the flow given by ∇W_λ , and diffeomorphic to l^∞.

The first way to exploit this situation would be to prove that rational G^λ_{pq}'s continue to some invariant set G^λ_ω in Y_ω, in the sense of Conley. Critical points in G^λ_ω would automatically have rotation vector ω, as elements of Y_ω. There the main stumbling block is: even though $\mathbf{Y} \subset (\mathbf{R}^n)$ can be endowed with the product topology, for which ∇L_λ is continuous, the "slices" Y_ω are not closed in that topology and hence do not constitute a local product parametrization for the flow ∇L_λ (with parameter $\omega \in \mathbf{R}^n$: for a definition of local product parametrization, i.e. the setting in which Conley's continuation is defined, see [C],[Sa]).

Another way to exploit this setting is the following. When we consider more restrictive boundary conditions ($R = 0$ outside a bounded strip), we can find in Y_ω a set P which has all the features of an isolating block except for compactness. Furthermore, P is the intersection of a compact set in $\mathbf{R}^{\mathbf{Z}}$ with Y_ω, it is a retraction of Y_ω, and finally it contains all the critical points in Y_ω (if there are any) and the critical torus corresponding to the KAM torus of rotation vector ω (when it exists). This gives us hope that some version of a continuation theorem may occur in Y_ω:

CONJECTURE. There exists a ghost torus in Y_ω, for each ω in $\mathbb{R}^n$

By ghost torus here we mean a compact invariant set whose cohomology contains the one of $\mathbb{T}^n$. Note that , supposing one could settle this conjecture positively, the problem of finding critical points would not be solved: the flow is not automaticaly gradiant like on Y_ω, making it hard to use the Conley-Zehnder Morse inequalities to find critical points. However, we think it would be a first step in finding critical points.

We should note here the results of Mather [Ma 2] on existence and regularity of minimal invariant measures for positive definite Lagrangian systems. Katok [K 3] proves that, for perturbations of completely integrable convex monotone maps there are infinitely many rotation vectors for which KAM tori do not exist but minimal orbits exist. Finally, Chen, McKay and Meiss [C-McK-M] found a whole class of (convex, monotone) 4D symplectic maps which have, for each irrational vector in a domain, cantori on which the motion is semi- conjugated to a rigid translation by that vector. These maps are perturbation of a discontinuous (sawtooth) map and are very far from completely integrable.

In the next section, we concentrate on twist maps ($n = 1$) and prove the conjecture in that case.

5. Ghost circles and twist maps. In this section, we concentrate on Twist maps and in particular answer positively to the conjecture of the last section for the case $n - 1$. When $n = 1$, the energy flow ζ^t has the remarkable property of being monotone with respect to the partial order on sequences. This fact was noticed by Angenent [An], whose work influenced the one exposed in this section a lot.

Let us recall a few facts about monotone flows (see, e.g. [Hi], [Mto]).

Let X be a Banach space. A *partial order* on X is given by a convex, closed cone V_+ such that, if we denote by $-V_+ = V_-$, we have: $V_- \cap V_+ = \{0\}$. For x, y in X we define:

$$\begin{aligned} x \leq y \quad &\Leftrightarrow \quad y \in x + V_+ \stackrel{\text{def}}{=} V_+(x) \\ x < y &\Leftrightarrow x \leq y \text{ and } x \neq y \\ [x, y] &= \{z \in X \quad | \quad x < z < y\} \\ (x, y) &= \{z \in X \quad | \quad x < z < y x \neq z \neq y\} \end{aligned}$$

DEFINITION 5. A map $A : X \to A$ is called *monotone* if:

$$x < y \Rightarrow Ax < Ay$$

A flow ζ^t in X is monotone if for all positive t, the map ζ^t is monotone.

One can also define the notion of *strong* monotonicity by replacing the cone V_+ by its interior in the above definition (the corresponding order is then denoted by $\ll$) . $\mathbb{R}^{\mathbb{Z}}$ is endowed with the natural partial order on sequences defined by the following: Let $\mathbf{x}$ and $\mathbf{x}'$ be elements of $\mathbb{R}^{\mathbb{Z}}$, with kth terms denoted by x_k and x'_k. Then:

$$V_+ := \{\mathbf{y} \in \mathbb{R}^{\mathbb{Z}} \quad | \quad y_k \geq 0, \forall k \in \mathbb{Z}\}.$$

and similarily for $<$.

The order in $X_{p,q}$ is induced by the one on $\mathbb{R}^{\mathbb{Z}}$. The positive cone is just the positive quadrant of $\mathbb{R}^q \cong X_{p,q}$. we also use the notation:

$$V_{++} := \{\mathbf{y} \in \mathbb{R}^{\mathbb{Z}} \quad | \quad y_k > 0, \forall k \in \mathbb{Z}\},$$
$$\mathbf{x} \prec \mathbf{x}' \Leftrightarrow \mathbf{x}' \in \mathbf{x} + V_{++}$$

and we will say that a map $A : \mathbb{R}^{\mathbb{Z}} \to \mathbb{R}^{\mathbb{Z}}$ is *strictly monotone* when

$$\mathbf{x} < \mathbf{x}' \Rightarrow A\mathbf{x} \prec A\mathbf{x}'.$$

It is clear that in $X_{p,q}$ the notion of strong monotonicity and strict monotonicity are equivalent.

LEMMA 1. *The energy flow is strongly monotone in $\mathbb{R}^{\mathbb{Z}}$ (and hence in $X_{p,q}$).*

In [Go 3] (Lemma 1.22), we present a proof of this which was communicated to us by S. Angenent. It does not rely on the classical theorem of Kamke, but rather proves that the solution operator of the linearised equation is strictly positive.

Remember that $\mathbb{Z}$ acts on $\mathbb{R}^{\mathbb{Z}}$ in two ways:

$$T, \sigma : \mathbb{R}^{\mathbb{Z}} \to \mathbb{R}^{\mathbb{Z}}$$
$$\{\sigma \mathbf{x}\}_k = x_{k+1}$$
$$\{T\mathbf{x}\}_k = x_k + 1$$

We will sometimes refer to the action of T as the $\mathbb{Z}$ action, to σ as the shift. There are natural projections of $\mathbb{R}^{\mathbb{Z}}$ into $\mathbb{A} \cong \mathbb{R}^2$. We denote by:

$$\pi_{01}(\mathbf{x}) = (x_0, x_1)$$

The results in this section concern the following objects:

DEFINITIONS 6 (σ -AUBRY-MATHER SETS, GHOST CIRCLES).

(1) A *σ-Aubry-Mather* set (or *σAM* set) is a closed, completely ordered, $\mathbb{Z}$-invariant, σ-invariant subset of $\mathbb{R}^{\mathbb{Z}}$.

(2) An *Aubry-Mather* set is a *completely critical σAM* set. We will say that an Aubry-Mather set is *saturated* if it is not strictly contained inside another one.

(3) A *ghost circle* is a ζ^t-invariant, connected σAM set.

COMMENTS. σAM sets are in a sense all the potential Aubry-Mather sets for all twist maps. The definition that we give here of an Aubry-Mather set is equivalent to the one of a F-invariant, T-invariant closed subset of $\mathbb{A}$ on which F preserves the order in the x coordinate and on which the first projection π_x is injective. (see [K1]).

The ghost circles as defined here are more than the $n = 1$ case of the ghost tori defined in the previous section. They are contained in the latter, sometimes strictly: consider the case when there exists a non ordered orbit of some rotation number. F-invariant circle are seen in $\mathbb{R}^{\mathbb{Z}}$ as completely critical ghost circles.

In [Go 3], we prove the following theorems of existence:

THEOREM 3 (EXISTENCE OF GHOST CIRCLES). *Let f be any twist map of $\mathbb{S}^1 \times \mathbb{R}$. Then for all ω in $\mathbb{R}$ there is a ghost circle G_ω with intrinseque rotation number $R(G_\omega) = \omega$.*

Furthermore, any Aubry-Mather set can be embedded into a ghost circle.

These ghost circles project diffeomorphically into the annulus to circles which are graphs over the x-axis and such that their images by f are also graphs. Intersections of such a circle and its image only occur at the critical points

The projection mentioned is given by $\mathbf{x} \to (x_0, x_1) \to (x_0, y_0)$ the last arrow being given by the diffeomorphism (3) in the definition of monotone maps. See theorem 5 for the significance of R.

THEOREM 4 (C^1 RATIONAL GHOST CIRCLES). *Suppose that the function $W_{p,q}$ attached to a twist map is Morse (proven to be a generic property in [Go 3]). Then:*

(1) There exists a ghost circle in $WO_{p,q}$ containing all absolute minima of $W_{p,q}$ and made of unstable manifolds of mountain pass points between them. These unstable manifolds join C^1 at the minima.

2) Alternatively, one can construct a ghost circle containing the extended orbits of an absolute minimum and its minimax. This ghost circle is also made out of unstable manifolds of mountain passes.

(3) In both cases, these ghost circles project diffeomorphically in the annulus to C^1 circles which are graphs over the x-axis and such that their images by f are also graphs. Intersections of such a circle and its image only occur at the periodic points.

The proof of theorem 3 derives immediately from a recent generalization of a theorem of Matano [Mto] by Dancer and Hess [D-H] on monotone flows in Banach spaces. This theorem states the existence of a monotone connecting orbit between two critical points $\mathbf{x} < \mathbf{y}$ whenever the order interval $(\mathbf{x}, \mathbf{y})$ does not contain any other critical points. The idea to apply their theorem is to fill the gaps in a saturated Aubry-Mather set with monotone connections. The σ and T invariance is given by the equivariance of the flow under these maps.

For theorem 4, we work in $X_{p,q}$ where we saturate the minimum periodic orbit with local minima that are ordered with it. We prove that between any two critical points in this order saturated set, there must be a mountain pass point of index 1 (note that this differs from Mather [Ma1] moutain pass lemma in that the extreme points of our order interval are only assumed to be *local* minima). We use properties of the unstable manifold of such points studied by Angenent [An] to prove that they join C^1 at the minima.

As far as the σAM sets are concerned, they provide the appropriate setting for the study of Hausdorff limits, as well as the rotation number. Remember that the rotation number of a sequence $\mathbf{x}$ in $\mathbb{R}^{\mathbb{Z}}$ is defined by:

$$\rho(\mathbf{x}) = \lim_{k \to \pm\infty} \frac{x_k}{k}$$

THEOREM 5.

(1) *If E is a connected σAM set, then it is homeomorphic to $\mathbb{R}$ and $E/T = E/\mathbb{Z}$ is homeomorphic to a circle. σ induces a circle homeomorphism on E/T.*

(2) *Any σAM set can be embedded in a connected σAM set and therefore is homeomorphic to a closed invariant set of a circle homeomorphism.*

(3) *The rotation number $R(E)$ of $\sigma \mid_E$ is a continuous function of E in the set of σAM sets.*

(4) *Let $\mathbf{x}$ be any element of E. Then $\rho(\mathbf{x})$ exists and $R(E) = \rho(\mathbf{x})$ for all $\mathbf{x}$ in E. Moreover, if $R(E) = \omega$, then*

$$E \subset Y_\omega \stackrel{\text{def}}{=} \{\mathbf{x} \in \mathbb{R}^{\mathbb{Z}} \quad | \quad \sup_{k\in\mathbb{Z}} |x_k - k\omega| < \infty\}$$

The last point of theorem 5 together with theorem 2 proves the conjecture of the last section in the case $n = 1$: the maximum compact invariant set in Y_ω actually contains a circle. It would be interesting to find out if the ghost tori that we find in higher dimensions contain actual tori.

6. Applications of ghost circles. As a first application, we have a formula for the flux of the map F through the projection G' onto the annulus of a ghost circle G. This flux is defined as the area above G' which is also below $F(G')$. The point being that this is well defined since both these sets are graphs over the x coordinate. Consider the set of connections between critical points in G and let K be the quotient of this set by the actions of σ and T. let $\underline{\mathbf{x}}(t)$ denote the class of a connection $\mathbf{x}(t)$ under this quotient. Letting $\lim_{t\to\pm\infty} \mathbf{x}(t) = \mathbf{x}^\pm$ one gets:

$$Flux(G') = \frac{1}{2} \sum_{\underline{\mathbf{x}}(t)\in E} \left| \sum_{k\in\mathbb{Z}} S(x_k^+, x_{k+1}^+) - S(x_k^-, x_{k+1}^-) \right|$$

Because of its geometric interpretation, the above sum converges. One can get rid of the factor 1/2 and the absolute value by chosing only the elements of E that give us positive terms. One can see the flux as the global variation of the energy W on G (eventhough W is not necessarily well defined there). When there is a direct connection $\mathbf{x}(t)$ between a min and a minimax, the sum corresponding to $\mathbf{x}(t)$ is the ΔW of Mather [Ma 1].

Let us remark that one can compute this flux by just knowing the critical points and that the setting is global, i.e. F can be as far as we want from completely integrable. We hope that this can become a valuable tool in the theory of transport. Could the higher dimensional ghost tori be used for this purpose as well?

Our second application offers a new setting for the study of the ΔW. We now look at ΔW as a function on the set of rational ghost circles:

DEFINITION 7. Let G_{pq} be a ghost circle in $X_{p,q}$ for a given twist map f. Then:

$$\Delta W(G_{pq}) = \max_{\mathbf{x},\mathbf{x}'\in G_{pq}} |W_{p,q}(\mathbf{x}) - W_{p,q}(\mathbf{x}')|$$

This is related to the definition of Mather by:

$$\Delta W_{p,q} \leq \Delta W(G_{pq}).$$

Consider a converging sequence of ghost circles $\{G_{p_k q_k}\}_{k\in\mathbf{Z}}$ with

$$G_{p_k q_k} \to G_\omega \text{ and } p_k/q_k \to \omega,$$

Then we have the following:

THEOREM 6.

(1) *If* $\lim_{k\to\infty} \Delta W(G_{p_k q_k}) = 0$ *then* G_ω *is a completely critical ghost circle (and hence projects to an* F*-invariant circle in* $\mathbb{A}$*).*

(2) *Conversally, if* G_ω *is completely critical then any sequence of rational ghost circles* $G_{p_k q_k}$ *with* $G_{p_k q_k} \to G_\omega$ *will satisfy:* $\lim_{k\to\infty} \Delta W(G_{p_k q_k}) = 0$

This inspired by the criterion of nonexistence of invariant circles of Mather [Ma], see also [K2].

If G_ω is transitive, then it is unique with this rotation number. Then any sequence $G_{p_k q_k}$ with $p_k/q_k \to \omega$ has $\Delta W(G_{p_k q_k}) \to 0$. Another way to say this is the following criterion:

If $G_{p_k q_k}$ *is a converging sequence of ghost circles and if* $\Delta W(G_{p_k q_k}) \not\to 0$*, then there is no transitive invariant circles of rotation number* $\omega = \lim_{k\to\infty} p_k/q_k$.

We could refine this criterion by erasing "transitive" in the above criterion if we could prove the unicity of G_ω when it is a completely critical set, not necessarily transitive. Also, we think that ΔW can actually be defined on σAM sets and that it is continuous there. The discontinuity at rational values of the ΔW of Mather would arise from the non unicity of the rational ghost circles.

REFERENCES

[An] S.B. ANGENENT, *The Periodic orbits of an area preserving twist map*, Comm. in Math. Physics, 115, no3 (1988).

[Ar] V.I. ARNOLD, *Mathematical Methods of Classical Mechanics*, Springer-Verlag, 1978.

[A-L] S. AUBRY AND P.Y. LEDAERON, *The discrete Frenkel-Kontorova model and its extensions. I. Exact results for ground states*, Physica 8D (1983), pp. 381-422.

[B-K] D. BERNSTEIN AND A.B. KATOK, *Birkhoff periodic orbits for small perturbations of completely integrable Hamiltonian systems with convex Hamiltonians*, Invent. Math., 88 (1987), pp. 225-241.

[Ch] A. CHENCINER, *La dynamique au voisinage d'un point elliptique conservatif, séminaire Bourbaki no. 622*, Asterisque, 121-122 (1985), pp. 147-170.

[C-McK-M] Q. CHEN, R.S. MCKAY AND J.D.MEISS, *Cantori for Symplectic maps*, submitted to Nonlinearity.

[Co] C.C. CONLEY, *Isolated invariant sets and the Morse index*, CBMS regional conference series in Math., 1978.

[C-Z 1] C.C. CONLEY AND E. ZEHNDER, *The Birkhoff-Lewis fixed point theorem and a conjecture of V.I Arnold*, Invent. Math. (1983), pp..

[C-Z 2] ———————, *Morse type index theory for Hamiltonian equations*, Comm. Pure and Appl. Math., XXXVII (1984), pp. 207-253.

[D-H] E.N. Dancer and P.Hess, *Stability of fixed points for order preserving discrete time dynamical systems*, preprint, University of New England, Australia, (1989).

[Fl1] A. Floer, *A refinement of the Conley index and an application to the stability of hyperbolic invariant sets*, preprint, University of Bohn.

[F-Z] A. Floer and E. Zehnder, *Fixed points results for symplectic maps related to the Arnold conjecture*, Collection: Dynamical systems and bifurcations (Groningen, 1984), Springer, Berlin-New York.

[Go 1] C. Golé, *Periodic points for monotone symplectomorphisms of* $\mathbf{T}^n \times \mathbf{R}^n$, Ph.D. Thesis, Boston University (1989).

[Go 2] ———, *Monotone maps of* $\mathbf{T}^n \times \mathbf{R}^n$ *and their periodic orbits*, to appear in proceedings of hamiltonian systems, MSRI.

[Go 3] ———, *Ghost circles for twist maps*, IMA preprint.

[G-H] C. Golé and G.R. Hall, *Poincaré's proof of Poincaré's last geometric theorem*, To appear in the proceedings for the workshop on twist maps, IMA, Springer.

[Ha] G. R. Hall, *A topological version of a theorem of Mather on twist maps*, Ergod. Th. and Dynam. Sys., 4 (1984), pp. 585-603.

[He] *Inegalités a priori pour des tores lagrangiens invariants par des difféomorphismes symplectiques*, Preprint, Ecole Polytechnique, (1989).

[Hi] M.W. Hirsch, *Stability and convergence in strongly monotone dynamical systems*, J. reine angew. Math., 383 (1988), pp. 1-53.

[J] F.W. Josellis, *Forced oscillations on* $\mathbf{T}^n \times \mathbf{R}^n$ *having a prescribed rotation vector*, preprint, Aachen (1987).

[K 1] A. Katok, *Some remarks on Birkhoff and Mather twist map theorems,*, Ergod. th. & dynam. sys., 2 (1982), pp. 185-194.

[K 2] ———, *More about Birkhoff periodic orbits and Mather sets for twist maps*, preprint, University of Maryland (1982).

[K 3] ———, *minimal orbits for small perturbations of completely integrable Hamiltonian systems*, preprint, California Institue of Technology.

[K-M] H. Kook and J.D. Meiss, *periodic orbits for reversible, symplectic mappings*, Phys.D, 35 (1989), pp. 65-86.

[McK-M-P] R.S.McKay, J.D. Meiss and I.I.C. Percival, *Stochasticity and transport in Hamiltonian systems*, Phys. Rev. Lett., 52 (1984), pp. 697-700.

[Mto] H. Matano, *Existence of nontrivial unstable sets for equilibriums of strongly order preserving systems*, J. Fac. Sci., U. Kyoto (1983), pp. 645-673.

[Ma 1] J.N. Mather, *A criterion for the non-existence of invariant circles*, Publ. Math. I.H.E.S., 63 (1986), pp. 153-204.

[Ma 2] ———, *Action minimizing invariant measures for positive definite Lagrangian systems*, preprint, ETH , Zurich.

[Mo 1] J.Moser, *Monotone twist mappings and the calculus of variation*, Ergod. th. and dyn. sys., 6 (1986), pp. 401-413.

[Mo 2] J.Moser, *The Birkhoff-Lewis fixed point theorem (appendix 3.3)*, appendix 3.3 in Lectures on closed geodesics , by W.Klingenberg, Springer.

POINCARÉ'S PROOF OF POINCARÉ'S LAST GEOMETRIC THEOREM

CHRISTOPHE GOLÉ* AND GLEN R. HALL†

Introduction. In this note we present a proof of the fixed point theorem known as Poincaré's Last Geometric Theorem. This theorem states that area preserving diffeomorphisms of the annulus which rotate the outer boundary clockwise and the inner boundary counter-clockwise must have a fixed point. It continues to play an important role in the study of Hamiltonian systems, differential equations and diffeomorphisms in spaces of low dimension (see [F2]).

Since Poincaré conjectured this theorem in his paper "Sur un théorème de géométrie" in 1912 [P], there have been many different proofs from the first, given by Birkhoff in 1913 [B1] to recent work of John Franks [F1]. Franks' version of the theorem greatly reduces both the geometric "twist" and the analytic area preservation hypotheses to obtain a theorem much stronger than Poincaré's original conjecture.

The purpose of this note is to review the original work of Poincaré in his 1912 paper. This contains not only the conjecture, but the proof of Poincaré's theorem in some special cases. The ideas used by Birkhoff and subsequent authors are considerably different from those of Poincaré (except for the attempt of Dantzig, see Birkhoff [B2]). In this note we show that Poincaré's original ideas can be modified slightly to yield a proof of his conjecture and give some interesting insights into the dynamics of these maps. The version of the theorem which we prove here is much weaker than Franks' theorem. We emphasize that this is **not** a correction of Poincaré's work – Poincaré was very clear on what he had and had not proven. This is rather an exposition and slight extension of Poincaré's ideas.

The authors would like to thank all those who listened patiently and assisted generously in this work, particularly Danny Goroff. The second author would also like to thank the Mathematics Department of the University of Cincinnati for its hospitality.

Definitions, notations and statement of the theorem. We let

$$\begin{aligned}
&\mathcal{A} = S^1 \times [0,1],\\
&A = \mathbf{R} \times [0,1],\\
&\pi : A \to \mathcal{A} : (x,y) \to (2\pi(x \mod 1), y),\\
&\left.\begin{matrix}\pi_x\\ \pi_y\end{matrix}\right\} : A \to \mathbf{R} : (x,y) \to \begin{cases} x \\ y \end{cases}.
\end{aligned}$$

denote the annulus, its universal cover the strip and the natural projection maps, respectively.

Notation : For $\tilde{f} : \mathcal{A} \to \mathcal{A}$ a diffeomorphism, we denote a (choice of) lift for $\tilde{f}$ by $f : A \to A$, so $\forall (x,y) \in A, f(x+1,y) = f(x,y) + (1,0)$.

*Department of Mathematics, State University of New York at Stony Brook, Stony Brook, New York 11794-3651.

†Department of Mathematics, Boston University, MA 22135.

DEFINITION. A diffeomorphism $\tilde{f} : \mathcal{A} \to \mathcal{A}$, or its lift $f : A \to A$ will be called a **twist map** if

(1) $\tilde{f}$ is isotopic to the identity,

(2) $\forall x \in \mathbf{R}, \pi_x(f(x,0)) < x$ and $\pi_x(f(x,1)) > x$,

(3) $\tilde{f}$ preserves an absolutely continuous finite invariant measure with support all of $\mathcal{A}$.

Remarks. 1) We have included condition (3) in the above definition to shorten the phrase "area preserving twist map" to "twist map". Also condition (2) is sometimes called the "boundary twist condition".

2) There may be several choices of the lift f which satisfy condition (2) (as well as infinitely many choices for which it is not satisfied). We assume that both $\tilde{f}$ and f are specified.

3) The smoothness of f is not important – the theorems below hold with f a homeomorphism. (e.g. see Franks [F2])

4) We could also work on the infinite cylinder (or its cover $\mathbf{R}^2$) but the boundary curves simplify exposition greatly.

Poincaré's last geometric theorem. Suppose $\tilde{f} : \mathcal{A} \to \mathcal{A}$ with lift $f : A \to A$ is a twist map. Then f has a fixed point.

Remarks. This is the theorem of Birkhoff (1913 [B1]). Birkhoff later improved this theorem, showing that such an $\tilde{f}$ has at least two fixed points ([B2], see also [BN]) and recent work of Franks [F2] have greatly weakened all of these hypotheses (see above).

Poincaré's idea in [P] was to study the set of points in A whose y-coordinate was not changed by the given twist map f. His goal was to show that if f has no fixed points then this set could be used to construct a loop in $\mathcal{A}$ which was contained in the exterior or interior of its image under $\tilde{f}$. Such a loop contradicts the area preservation hypothesis and hence every twist map must have a fixed point. Hence we need to develop some machinery for constructing the required curve. The following section developes the topological machinery. Next the rules for the (inductive) construction of the required loop are described.

Topological preliminaries and lemmas. We will use the following:

DEFINITION. If $\gamma : S^1 \to \mathbf{R}^2$ is a continuous map, for each point $z \in \mathbf{R}^2 \sim \gamma(S^1)$ we define the index of z to be the degree of the map $S^1 \to S^1 : t \to \frac{\gamma(t)-z}{||\gamma(t)-z||}$ where $||\cdot||$ is the usual $\mathbf{R}^2$ norm.

Remark. This is the usual winding number of a loop in $\mathbf{R}^2$ (hence a continuous, integer valued function on $\mathbf{R}^2 \sim \gamma(S^1)$). For definiteness we choose orientations so that a counterclockwise circle has index $+1$ about points in its interior.

DEFINITION. Let $\gamma_1, \gamma_2 : [0,1] \to \mathbf{R}^2$ be two continuous arcs and let $\gamma_1 - \gamma_2$ denote the map $\gamma_1 - \gamma_2 : [0,4] \to \mathbf{R}^2$ given by

$$t \to \begin{cases} \gamma_1(t) & \text{if } 0 \le t \le 1 \\ (2-t)\gamma_1(1) + (t-1)\gamma_2(1) & \text{if } 1 \le t \le 2 \\ \gamma_2(3-t) & \text{if } 2 \le t \le 3 \\ (4-t)\gamma_2(0) + (t-3)\gamma_1(0) & \text{if } 3 \le t \le 4 \end{cases}$$

We say $\gamma_1 - \gamma_2$ has *positive index* if it has positive or zero index about every point of $\mathbf{R}^2 \sim (\gamma_1 - \gamma_2)([0,4])$. (See Figure 1).

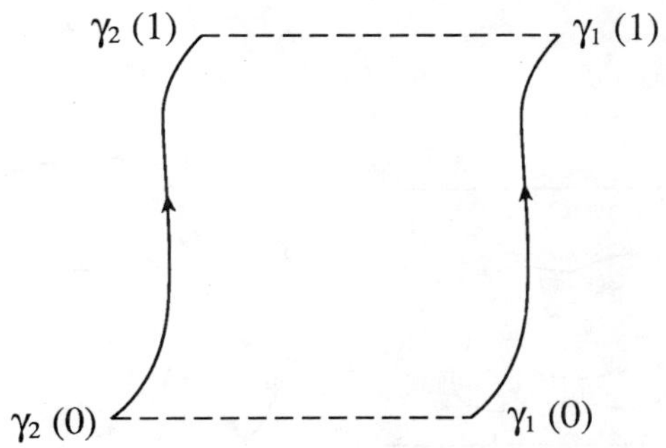

Figure 1. $\gamma_1 - \gamma_2$ has positive index.

Notation : For $\gamma, \delta : S^1 \to \mathbf{R}^2$ simple closed loops, we say $\gamma - \delta$ *has positive index* if when we consider $\gamma, \delta : [0, 2\pi] \to \mathbf{R}^2$ as arcs then $\gamma - \delta$ has positive index in the sense above. (This involves a rescaling of the domain of $[0, 2\pi]$ to $[0,1]$ in the obvious way).

LEMMA 1. *Suppose $\gamma, \delta : S^1 \to \mathbf{R}^2$ are simple closed loops. If there exist arcs $\gamma_i, \delta_i : [0,1] \to \mathbf{R}^2, i = 1, \ldots, n$ satisfying*

(1) $\forall i = 1, \ldots, n-1, \gamma_i(1) = \gamma_{i+1}(0), \delta_i(1) = \delta_{i+1}(0)$ *and* $\gamma_n(1) = \gamma_1(0), \delta_n(1) = \delta_1(0)$,

(2) $\cup_{i=1}^n \gamma_i([0,1]) = \gamma(S^1), \cup_{i=1}^n \delta_i([0,1]) = \delta(S^1)$,

(3) $\gamma_i - \delta_i$ *is positive index for* $i = 1, \ldots, n$,

then $\gamma - \delta$ has positive index.

Proof. This follows easily from the additivity properties of index. □

LEMMA 2. *Suppose $\gamma, \delta : S^1 \to \mathbf{R}^2$ are simple closed loops with γ and δ both having index $+1$ or zero about every point of $\mathbf{R}^2 \sim \gamma(S^1)$ or $\mathbf{R}^2 \sim \delta(S^1)$, respectively. Let $U = \{z \in \mathbf{R}^2 \sim \gamma(S^1) :$ the index about z is $+1\}$. If $\gamma - \delta$ is positive index than $\delta(S^1) \subseteq$ closure (U).*

Proof. Since this index of points with respect to $\gamma - \delta$ is the difference of their indices with respect to γ and with respect to δ, each point of positive index of δ

must also be a point of positive index of γ. Since δ is a simple closed curve, this implies $\delta(S^1) \subseteq$ closure (U). □

Poincaré's idea for the proof of Poincaré's last geometric theorem. Fix $\tilde{f} : \mathcal{A} \to \mathcal{A}$ a twist map with lift $f : A \to A$. Poincaré's idea was to assume that f has no fixed points, then construct a curve $L \subseteq A$ which is either a simple closed loop, or has $\pi(L)$ a simple closed loop in $\mathcal{A}$, such that $L - f(L)($ or $\pi(L) - \tilde{f}(\pi(L)))$ has positive index. This contradicts the assumption of area preservation for f because lemma (3) implies that the region "inside" $L($ or $\pi(L))$ will be mapped inside itself by f (or $\tilde{f}$). (see Figure 2).

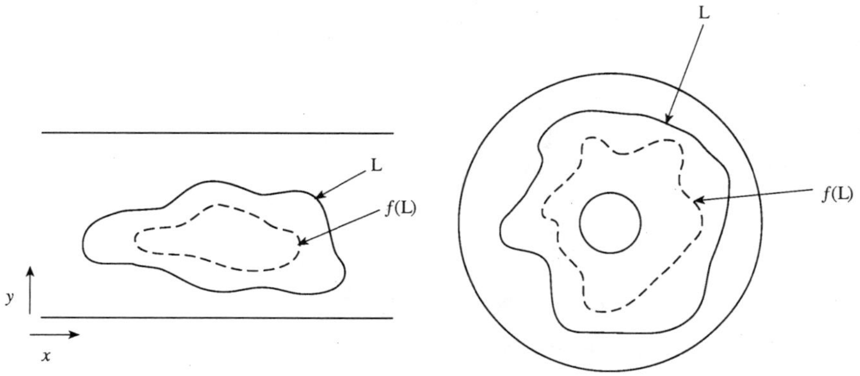

Figure 2.

To describe Poincaré's construction we need to define

$$\Gamma_f = \Gamma = \{z \in A : \pi_y(f(z)) = \pi_y(z)\}.$$

Typically, Γ will be a nice set. In fact, we have

LEMMA 3. *The twist maps $\tilde{f} : \mathcal{A} \to \mathcal{A}$ satisfying the following conditions are dense in the C^2 topology.*

(1) $\forall \varepsilon > 0, \Gamma_f \cap \{(x, y) \in A : \varepsilon < y < 1 - \varepsilon, 0 \leq x \leq 1\}$ *is an immersed one manifold with finitely many components.*

(2) $\forall \varepsilon > 0, \Gamma_f \cap \{(x, y) \in A : \varepsilon < y < 1 - \varepsilon, 0 \leq x \leq 1\}$ *has finitely many points with tangent parallel to* $(0, 1)$, *all with different y coordinate,*

(3) $\forall \varepsilon > 0$, *the sign of* $\pi_y(f(z)) - \pi_y(z)$ *is different between any two adjacent components of* $\{(x, y) \in A : \varepsilon < y < 1 - \varepsilon\} \sim \Gamma_f$.

Proof of lemma 3. We recall a few facts from transversality theory. For more details, we refer to [DeM, P], chapter 1, of which the following is taken (replacing C^∞ by C^1).

Let Λ, M, N be manifolds and let $F : \Lambda \times M \to N$ be a C^1 map. For $\lambda \in \Lambda$, we denote by $F_\lambda : M \to N$ the map defined by $F_\lambda(p) = F(\lambda, p)$. Let $S \subset N$ be a C^1 submanifold and let $T_S \subset \Lambda$ be the set of points λ such that F_λ is transversal to S.

PROPOSITION (3.3 P.25 IN [DEM, P]). *If $F : \Lambda \times M \to N$ is transverse to $S \subset N$, then T_S is residual in Λ.*

We can replace C^∞ by C^1, by using the C^1 version of Sard's theorem which is the cornerstone of the proof of the proposition. To apply this proposition in our situation, we will consider the map:

$$F : \mathbf{R}^2 \times A_\epsilon \to \mathbf{R} \quad \text{defined by}$$
$$F(\lambda, \mu, x, y) = \pi_y(g_{\lambda\mu}(x, y)) - y$$

where $A_\epsilon = \{(x, y) \in A \mid \epsilon < y < 1 - \epsilon, 0 \le x \le 1\} and g_{\lambda\mu}$ is the monotone twist map generated by the function $H_{\lambda\mu}(x, \bar{x}) = \frac{1}{2\epsilon}(\bar{x} - x)^2 + \frac{\lambda}{2\pi}\cos 2\pi x + \frac{\mu}{2\pi}\sin 2\pi x$ where λ and μ are $0(\epsilon^2)$, $H_{\lambda\mu}$ can be extended to all of A in such a way as to generate a monotone twist map $g_{\lambda,\mu}$ satisfying:

$$g_{\lambda\mu}|_{A_\epsilon}(x, y) = (x + \epsilon(y - \lambda\sin 2\pi x + \mu\cos 2\pi x), y - \lambda\sin 2\pi x + \mu\cos 2\pi x)$$
$$g_{\lambda\mu}|_{A^c_{\epsilon/2}}(x, y) = (x + \epsilon y, y)$$

hence $g_{\lambda\mu} \circ f$ is a twist map and $\|g_{\lambda\mu} \circ f - f\|_{c^2} \sim 0(\epsilon)$ (One continues the constant λ, μ as functions of $(\bar{x} - x)$ satisfying $\|\lambda\|_{C^2}$, $\|\mu\|_{C^2} \sim 0(\epsilon)$ and with graphs of the form:

Denote by $f_1(z) = \pi_x f(z), f_2(z) = \pi_y f(z)$. Then, on $\mathbf{R}^2 \times A_\epsilon$, we have:

$$F(\lambda, \mu, z) =$$
$$f_2(z) - \lambda\sin(2\pi f_1(z)) + \mu\cos(2\pi f_2(z)) - \pi_y(z)$$

We need to show that, for a generic set of λ, μ, F is transverse to the 0-dimensional manifold $\{0\}$, i.e., according to the proposition that DF is always onto $\mathbf{R}$.

But

$$\frac{\partial F}{\partial \lambda}(\lambda, \mu, z) = \quad \sin(2\pi f_1(z))$$
$$\frac{\partial F}{\partial \mu}(\lambda, \mu, z) = \quad \cos(2\pi f_1(z))$$

which can't be simultaneously 0.

Hence, for generic values of (λ, μ), $F^{-1}_{(\lambda,\mu)}(0)$ is a 1-dimensional manifold. The proofs of the other two statements are similar. For the second statement, one consider the 0 set of the map $A \to \mathbf{R}^2$ given by:

$$z \to (f_2(z) - \pi_y(z), f_2'(z)).$$

One needs 4 independent parameters to unfold the singularities. Take:

$$h_{\lambda\mu\bar{\lambda}\bar{\mu}} = \frac{1}{2\epsilon}(\bar{x} - x)^2 + \frac{\lambda}{2\bar{n}}\cos 2\pi x + \frac{\mu}{2\bar{u}}\sin 2\pi x$$
$$+ \frac{\bar{\lambda}}{4\bar{u}}\cos 4\pi x + \frac{\bar{\mu}}{4\bar{u}}\sin 4\pi x$$

we leave the details to the reader. □

Also, we can easily see that the structure of Γ is related to the fixed points of f by the following:

LEMMA 4. *Suppose $\tilde{f} : \mathcal{A} \to \mathcal{A}$ is a twist map. Then there exists an $\varepsilon > 0$ such that if Γ contains a (connected) component $\Gamma_0 \subseteq \Gamma_f$ which connects the line $y = \varepsilon$ to the line $y = 1 - \varepsilon$ then Γ_0 contains a fixed point of $\tilde{f}$.*

Proof. When $\varepsilon > 0$ is sufficiently small we have $\forall x \in \mathbf{R}, \pi_x(f(x, \varepsilon)) < x$ and $\pi_x(f(x, 1-\varepsilon)) > x$. But, if Γ_0 connects $y = \varepsilon$ to $y = 1-\varepsilon$ then for some $z_0, z_1 \in \Gamma_0$, we have $\pi_x(f(z_0)) < \pi_x(z_0)$ and $\pi_x(f(z_1)) > \pi_x(z_1)$. So, by continuity, there exists $z \in \Gamma_0$ with $\pi_x(f(z)) = \pi_x(z)$. But then z is the required fixed point because $z \in \Gamma_0$ implies $\pi_y(f(z)) = \pi_y(z)$, so $\tilde{f}(z) = z$. □

Now, the loop L above will be constructed from segments of Γ and horizontal (y = constant) line segments contained in lines tangent to Γ such that each segment $\ell \subseteq L$ has $\ell - f(\ell)$ positive index. Poincaré's method was to construct a large, oriented graph in A from segments of Γ and horizontal segments in lines tangent to Γ with edges oriented so that an edge minus its image under f has positive index. If this graph contains an oriented loop then this is the required loop L. Unfortunately, Poincaré could not show that his directed graph contained an oriented loop (although he worked out numerous examples which yielded the required loop).

He states that if there is no such loop, he could construct a counterexample to the theorem. (Hence, he showed the existence of the loop in his graph is equivalent to the fixed point theorem).

We alter the construction slightly, building the curve L by adding arcs ℓ_i inductively to a curve in A such that $\ell_i - f(\ell_i)$ has positive index and ℓ_i is either an arc of Γ or a horizontal segment on a line tangent to Γ with end points in Γ. When this curve closes (as it must by the generic finiteness assumption on Γ) in A or $\mathcal{A}$, the resulting curve will be the required loop. This is really just a slight modification of the last step of Poincaré's construction.

We note that if we can prove the theorem for maps satisfying the conditions of lemma 3, then the theorem for arbitrary twist maps follows by the usual limit arguments (i.e., the limit of maps on a compact space having fixed points will have a fixed point).

Construction of L. Fix a twist map $\tilde{f} : \mathcal{A} \to \mathcal{A}$ with lift $f : A \to A$ which satisfies the conditions in Lemma (3) (i.e., Γ is made of smooth arcs etc.). We assume, for contradiction, that f has no fixed points. We will construct the loop $L \in A$ discussed above by piecing together inductively arcs ℓ_i such that $\ell_i - f(\ell_i)$ are positive index. The arcs ℓ_i will be pieces of Γ or horizontal segments on lines tangent to Γ, so eventually this curve will form a loop in A or its projection will form a loop in $\mathcal{A}$. The steps in this construction are the following:

I) Orient and label the components of Γ and $A \sim \Gamma$,

II) Describe the rules for choosing ℓ_{n+1} given $\ell_0, \ldots, \ell_n$,

III) Prove that ℓ_{n+1} exists and $\ell_{n+1} - f(\ell_{n+1})$ has positive index,

IV) Show the construction yields the required loop.

Step I: First we label the components of $A \sim \Gamma$ either "up" , if $\pi_y(f(z)) > \pi_y(z)$ or "down", if $\pi_y(f(z)) < \pi_y(z)$ for z in the component.

Next we label components of Γ as either "left" if $\pi_x(f(z)) < \pi_x(z)$ or "right" if $\pi_x(f(z)) > \pi_x(z)$ for z in the component. To orient components of Γ we first attach to each point $z \in \Gamma$ a vector $\underline{n}_z$ normal to Γ such that if z is on a left component of Γ then $\underline{n}_z$ points into an up region of $A \sim \Gamma$ while if z is on a right component of Γ then $\underline{n}_z$ points into a down component of $A \sim \Gamma$. Now we orient components of Γ by choosing a vector tangent to Γ at $z \in \Gamma$, called $\underline{t}_z$, such that $(\underline{t}_z, \underline{n}_z)$ has the same orientation as the standard basis $((1,0),(0,1))$. (See Figure 3).

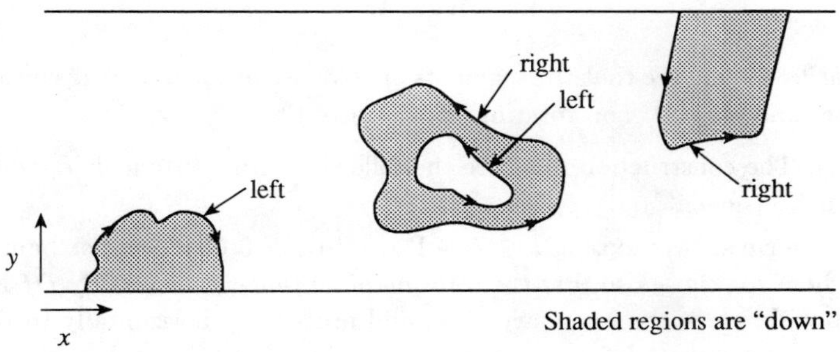

Figure 3.

Next we must break Γ into two disjoint subsets – the points of Γ whose tangent vectors are rotated by f a large amount and those for which the tangent vectors are rotated only a small amount. The fact that f preserves the y coordinate of Γ will allow us to distinguish "large" from "small" amounts of rotation as follows: To each non-zero tangent vector on A assign an angle given by the absolute value of angle between the vector and its image under the derivative of f. In order to remove the ambiguity mod 2π in this choice we require that the tangent vectors tangent to the boundary are assigned zero (they are preserved by the derivative of f, hence have angle $2n\pi$ for $n \in \mathbf{Z}$ – we choose $n = 0$) and we require the choice be continuous in A (hence some angles might be larger than 2π, etc.). Now we divide Γ into two pieces

$$\Gamma_N = \{z \in \Gamma : \text{ the tangent vector } \underline{t}_z \text{ is rotated by less than } \pi \text{ by the derivative of } f\}.$$

$$\Gamma_R = \{z \in \Gamma : \text{ the tangent vector } \underline{t}_z \text{ is rotated by } \pi \text{ or more by the derivative of } f\}.$$

Then $\Gamma = \Gamma_N \cup \Gamma_R$ and $\Gamma_N \cap \Gamma_R = \emptyset$. (See Figure 4.)

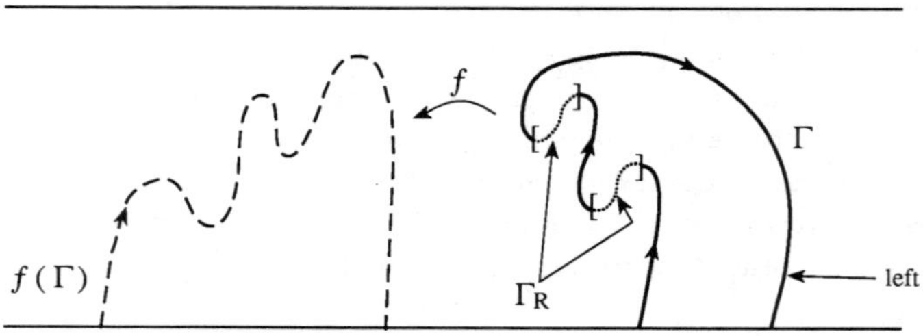

Figure 4.

Remark. We will see that the segments of Γ we use in the following construction are all contained in the "non rotating" part Γ_N of Γ.

Step II: The construction rules are the following: Rule (0) yields ℓ_0 while rules (1-3) yield ℓ_{n+1} given $\ell_0, \ldots, \ell_n$:

0) ℓ_0 is formed by choosing $z \in A \sim \Gamma$ within $\varepsilon > 0$ of the upper boundary of $A \sim \Gamma$ (where ε is chosen so that the rotation of all tangent vectors by Df is small) for points within ε of the boundary of A, and proceeding horizontally to the right if z is in an up region or left if z is in a down region of $A \sim \Gamma$ until a component of Γ (which must be in Γ_N) is encountered.

1) If n is even, so ℓ_n is a horizontal segment from end points z_0 in ℓ_{n-1} to z_1 in Γ, then ℓ_{n+1} is formed by following Γ from z_1 in the direction of the orientation assigned in step I until either a point of horizontal tangency is encountered or a "jump point" as described below in rule (2) is encountered.

2) If $n+1$ is odd, so ℓ_{n+1} is being formed by following a component of Γ as in rule (1), we identify two types of "jump points" as follows;

type 1: Suppose ℓ_{n+1} is on a left component of Γ, a point z of ℓ_{n+1} is called a jump point if there is a point $w \in \Gamma$ with

- •) w is on a left component of Γ,
- •) $\pi_y(z) = \pi_y(w), \pi_x(z) > \pi_x(w)$ and $\pi_x(f(z)) < \pi_x(f(w))$,
- •) if $\overline{zw}$ is the horizontal segment connecting z and w then $f(\overline{zw})$ is homotopic to $\overline{f(z)f(w)}$ rel $f(\ell_{n+1}) \cup \{f(w)\}$ (see Figure 5).

type 2: Suppose ℓ_{n+1} is on a right component of Γ, a point z of ℓ_{n+1} is called a jump point if there is a point $w \in \Gamma$ with

- •) w is on a right component of Γ,
- •) $\pi_y(z) = \pi_y(w)$, $\pi_x(z) < \pi_x(w)$ and $\pi_x(f(z)) > \pi_x(f(w))$,
- •) if $\overline{zw}$ is the horizontal segment connecting z and w then $f(\overline{zw})$ is homotopic to $\overline{f(z)f(w)}$ rel $f(\ell_{n+1}) \cup \{f(w)\}$, (see Figure 5).

If while constructing ℓ_{n+1} a jump point z is encountered then ℓ_{n+1} is continued by the segment $\overline{zw}$ then following Γ from w in the direction of the orientation of step I (then return to rule (1)). (see Figure 6).

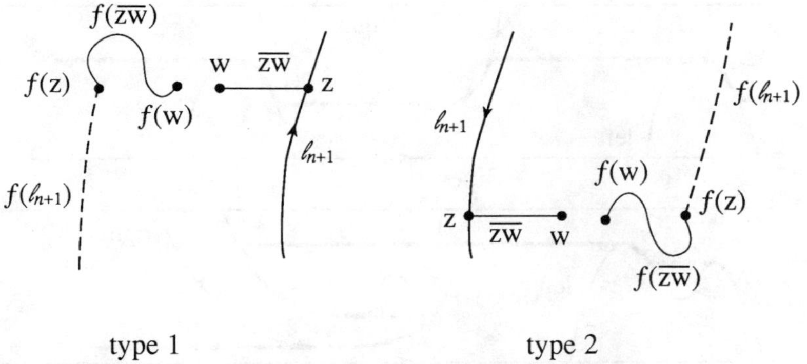

Figure 5.

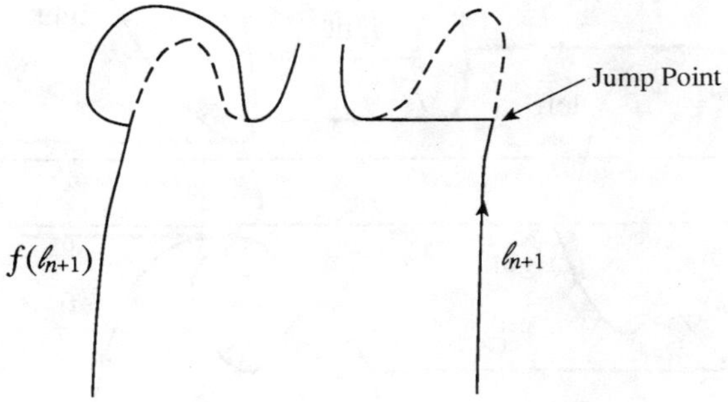

Figure 6.

3) If n is odd, so ℓ_n is made up of segments of Γ and horizontal segments from jump points and ℓ_n ends at a point z_0 of horizontal tangency of Γ, then ℓ_{n+1} is formed by the longest horizontal segment beginning at z_0 in the direction of the tangent to Γ at z_0 so that

i) $\ell_{n+1} - f(\ell_{n+1})$ has positive index,

ii) both end points of Γ are in Γ_N,

iii) neither end point is a jump point (as in rule (2)),

iv) segments of ℓ_{n+1} in $A \sim \Gamma$ nearest the end points are either both in up or both in down components of $A \sim \Gamma$,

We give the following examples in Figure 7. These should be compared with the figures in Poincaré [P], however, we warn the reader that the labelling of Γ is different here.

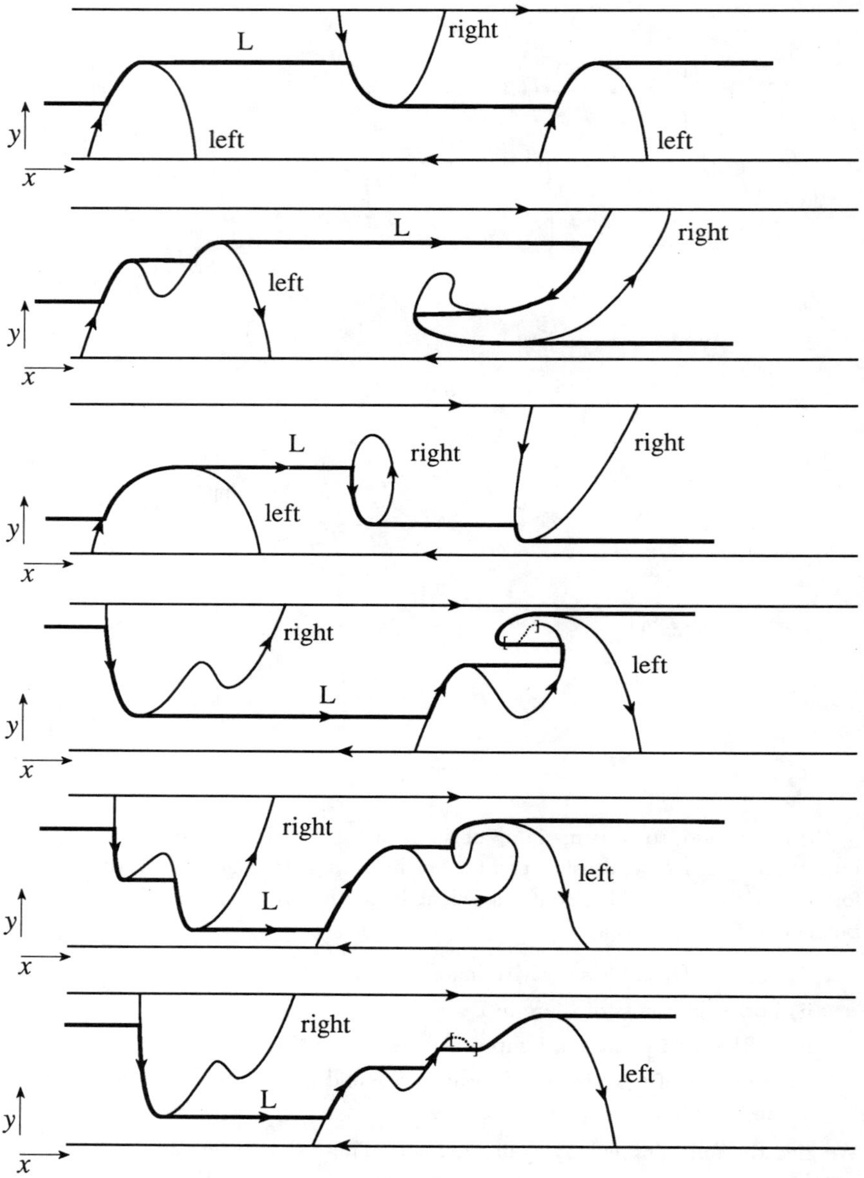

Note: [dotted bracket symbol] indicates pieces of Γ in Γ_R

Figure 7.

Step III: Next we must show that the rules above produce the desired curve. That is, we must show that there is always a horizontal segment satisfying the conditions of rule (3) and that the curves ℓ created by rules (1) and (2) have $\ell - f(\ell)$ positive index. We begin with two lemmas giving examples of positive index arcs.

LEMMA 5. *Suppose ℓ is a horizontal segment with end points $z_1, z_2 \in \Gamma$ and interior in $A \sim \Gamma$ oriented from z_1 to z_2. Then if the signs of $\pi_x(z_1) - \pi_x(z_2)$ and $\pi_x(f(z_1)) - \pi_x(f(z_2)$ agree and ℓ is in an up component of $A \sim \Gamma$ if $\pi_x(z_1) < \pi_x(z_2)$ or ℓ is in a down component of $A \sim \Gamma$ if $\pi_x(z_2) < \pi_x(z_1)$ then $\ell - f(\ell)$ has positive index.*

More generally, let γ be a curve in A satisfying the following:

(1) $\pi_y\gamma(0) = \pi_y\gamma(1)$ *and* $\pi_x\gamma(1) > \pi_x\gamma(0)$

(2) $\pi_y(\varepsilon) > \pi_y\gamma(0)$

(3) *γ is homotopic to the segment $\ell = \overline{\gamma(0)\gamma(1)}$ rel $\{\gamma(0), \gamma(1)\}$*

(4) *The interior of the lift of γ in the universal covering of $A \sim \{\gamma(0), \gamma(1)\}$ does not intersect the lift of $\overline{\gamma(0)\gamma(1)}$ to which it is homotopic.*

Then the curve $\ell - \gamma$ has positive index.

Remark. The types of curves γ allowed by the second portion of the lemma are precisely those specified by requiring $\ell - \gamma$ to be positive index. They are those for which the intersections of ℓ with γ occur with non-zero winding (see Figure 8).

This lemma seems to be a particular case of a more general property of Jordan curves, where one would consider whether such a curve self intersects "from inside" or "from outside". "Outside" intersections add index, "inside" ones can give negative index. We can tell the "inside" intersection from the "outside" by looking at lifted curves in te appropriate covering spaces as in the statement of the lemma.

Proof. This follows recalling that positive index is counterclockwise (see Figure 8). □

LEMMA 6. *Suppose $\ell \subseteq \Gamma$ is an arc with end points z_1 and z_2 oriented from z_1 to z_2 such that the interior of ℓ contains no points of horizontal tangency of Γ. If ℓ is on a left component of Γ and $\pi_y(z_1) < \pi_y(z_2)$ or if ℓ is on a right component of Γ and $\pi_y(z_1) > \pi_y(z_2)$ then $\ell - f(\ell)$ has positive index.*

Proof. Again, follows from the choice of orientation (see Figure 9.). □

In the next 3 lemmas, we show that with two additional assumptions, there will be horizontal segments satisfying the conditions of rule (3). Rules (3) i and iv are treated in lemma 7, rule ii (and more) in lemma 8 and lemma 9 deals with iii.

LEMMA 7. *Suppose n is odd so that ℓ_n contains components of Γ and ends at a point z_0 of horizontal tangency of Γ. Suppose also that*

a) *$z_0 \in \Gamma_N$, the non-rotating part of Γ.*

b) *if the tangent to Γ at z_0 points right then it points into an up region of $A \sim \Gamma$, if left then it points into a down region of $A \sim \Gamma$.*

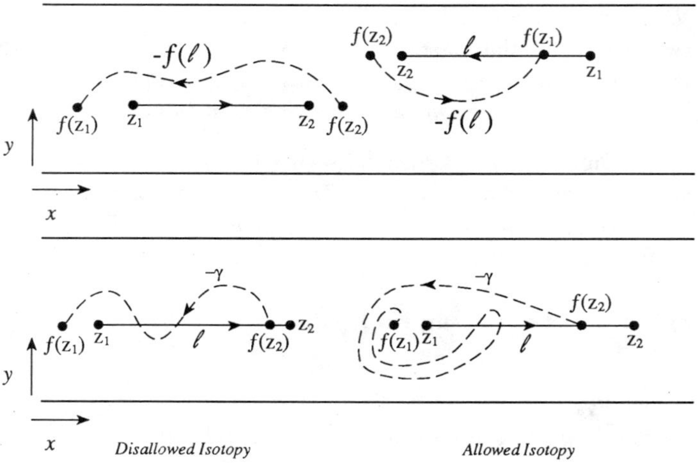

Figure 8.

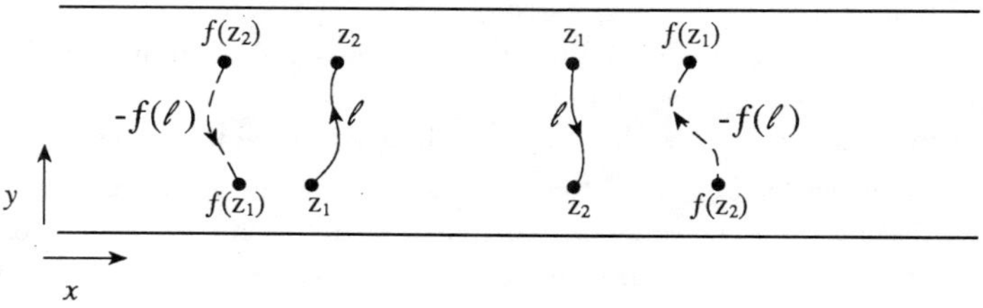

Figure 9.

Then there is a horizontal segment ℓ_{n+1} satisfying rule (3) i and iv.

Proof. Suppose the tangent to Γ at z_0 points to the right. Let R_0 be the horizontal ray beginning at z_0 and extending to the right and let R_1 be the horizontal ray starting at $f(z_0)$ and extending to the right.

Claim. By condition (a) of the Lemma, $f(R_0)$ will be homotopic to R_1 rel $f(\ell_n)$.

Proof of claim. (see Figure 10). If we form the suspension of f in $A \times [0,1]$ then we see that the points to the far right must stay to the right of ℓ_n and $f(\ell_n)$. Since the tangent vector to Γ at z_0 does not rotate and the image of R_0 must be as in the claim. □

Since $f(R_0)$ must cross R_1 infinitely many times, there must be a point $f(z_1)$ in $f(R_0) \cap R_1$ such that the points to the left of z_1 are in an up component and such that $f(\overline{z_0 z_1})$ is homotopic to $\overline{f(z_0)f(z_1)}$ rel $f(\ell_n) \cup f(z_1)$.

Condition (b) and lemma 5 complete the proof (See Figure 10.). The other case is symmetric. □

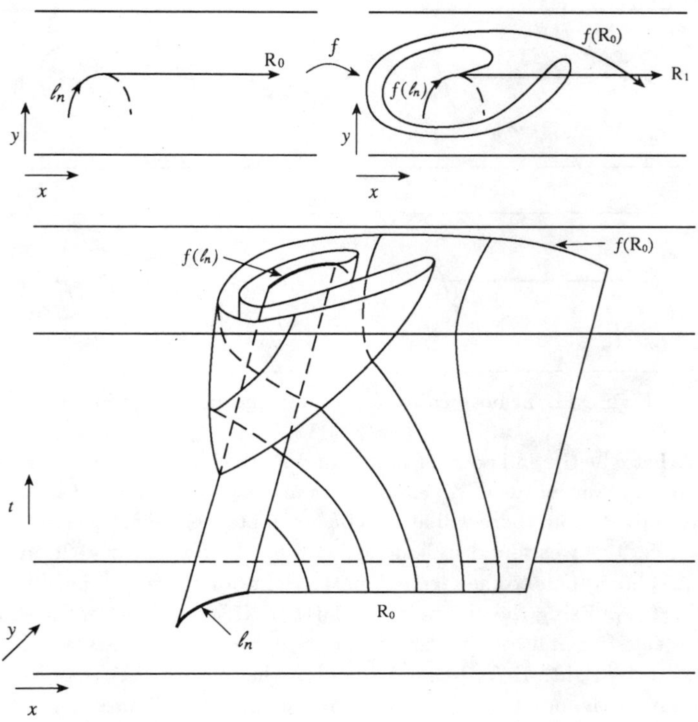

Figure 10.

Next we consider the type of components of Γ on which ℓ_{n+1} of the above lemma can end. In particular, we prove that ℓ_{n+1} satisfies condition ii of rule 3).

LEMMA 8. *Suppose n is odd and ℓ_n satisfies conditions a and b of lemma 7, then the end point $z_1 \in \Gamma$ of ℓ_{n+1} which is not in ℓ_n will be in Γ_N and if it is on a left (respectively, right) component of Γ then the tangent to Γ at z_1 will point up (respectively, down), i.e., will have positive, (respectively, negative) y component.*

Proof. Since the segments of ℓ_{n+1} nearest the end points are either both in up region if ℓ_{n+1} is oriented to the right or both in down regions if it is oriented to the left, the rules of orientation of Γ imply that left (respectively, right) component of Γ at the end of ℓ_{n+1} will be oriented upward (respectively, downward).

Since the beginning point of ℓ_{n+1} is in Γ_N by assumption, it follows that the end point must also be in Γ_N, since ending at a point in Γ_R would not satisfy $\ell_{n+1} - f(\ell_{n+1})$ having positive index, (see Figure 11). □

Finally, we must show that there is always a choice of ℓ_{n+1} satisfying the conditions of rule (3)iii.

LEMMA 9. *For n odd, assuming ℓ_n satisfies conditions a and b of lemma 7, there exists a horizontal segment ℓ_{n+1} satisfying the conditions of rule (3).*

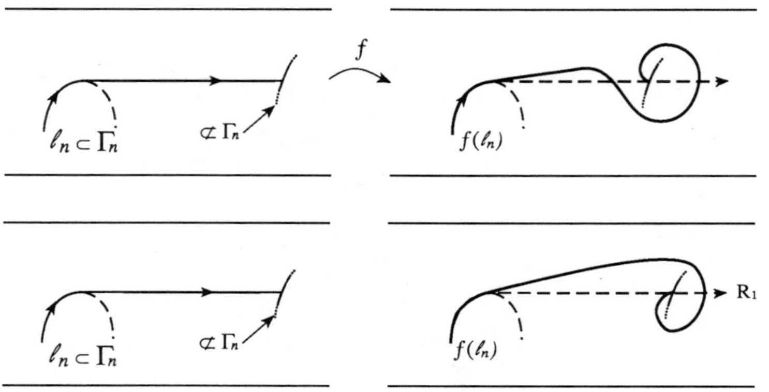

Figure 11. In both cases, ℓ_{n+1} must end in Γ_R not in Γ_R.

Proof. Let a be the end point of ℓ_n in Γ and R_0 the horizontal ray pointing in the direction of the tangent at a. We assume a is in a left component of Γ. Then since f preserves orientation, there will always be horizontal segments ℓ_{n+1} satisfying i,ii, and iv of rule 3 as was found in lemma 7 and 8. We note that such an ℓ_{n+1} will have $f(\ell_{n+1})$ homotopic to the segment on the right pointing ray connecting the end points of $f(\ell_{n+1})$ fixing the end points, rel $f(\ell_n)$. If such a segment is chosen and it does not satisfy condition iii, then it may be extended to satisfy all of i-iv when R_0 points to the right. If R_0 points to the left then either ℓ_{n+1} may be extended to satisfy i-iv or the point a is already a jump point. (See Figure 12). □

Next we show that for n odd, the arcs ℓ_n will end at points of Γ_N. (assumption (a) of lemma 7).

LEMMA 10. *Suppose n is odd and that ℓ_n has its last segment with interior in Γ_N and is either upward oriented left or downward oriented right component of Γ. Then the end point of ℓ_n at a point of horizontal tangency of Γ is in Γ_N.*

Proof. Suppose not. We assume ℓ_n is in an upward oriented left component of Γ. Then the local picture near the end of ℓ_n is as in Figure 13.

Now case (a) does not occur because the component of $A \sim \Gamma$ to the left and above ℓ_n must be up. Hence we need only consider case (b). Now we consider the continuation of Γ after ℓ_n. Since its image is trapped by the image of the horizontal ray, we see we must have Γ again crossing the line horizontally tangent to Γ as in Figure 14.

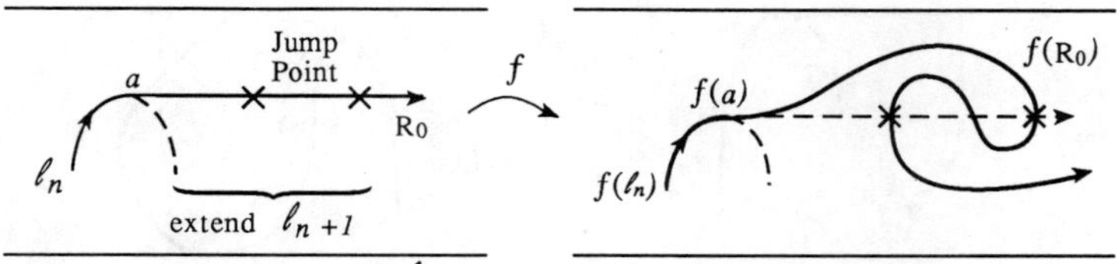

extend $\ell_n +1$ to satisfy *i-iv* of rule (3)

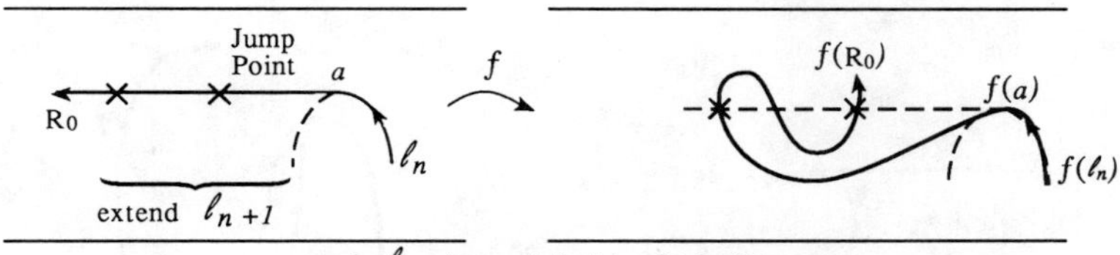

extend $\ell_n +1$ to satisfy *i-iv* of rule (3)

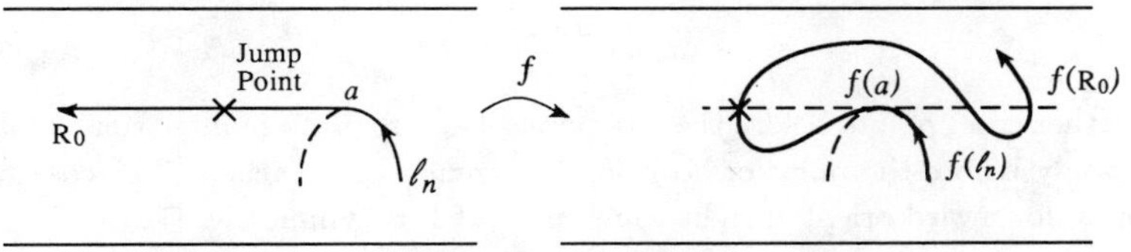

(*a*) is already a jump point!

Figure 12.

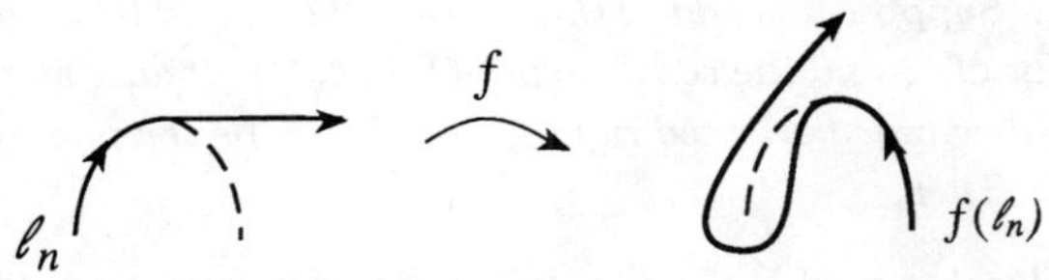

Case a

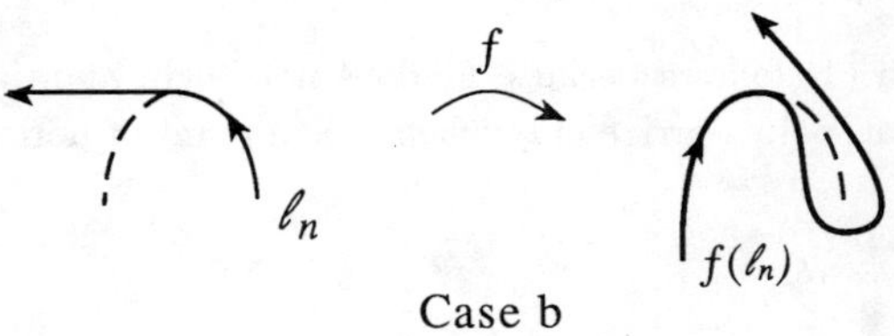

Case b

Figure 13.

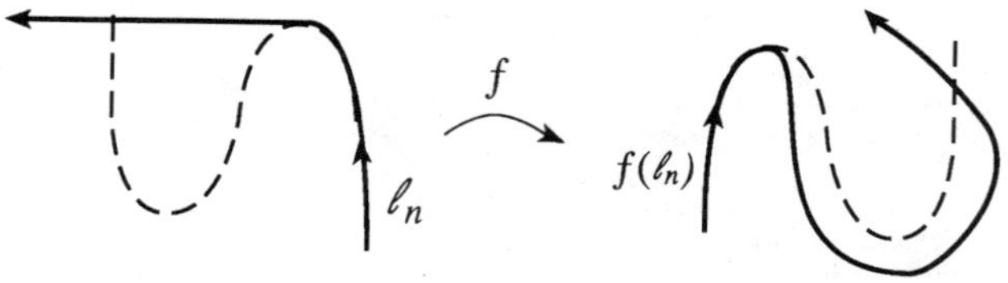

Case (i)

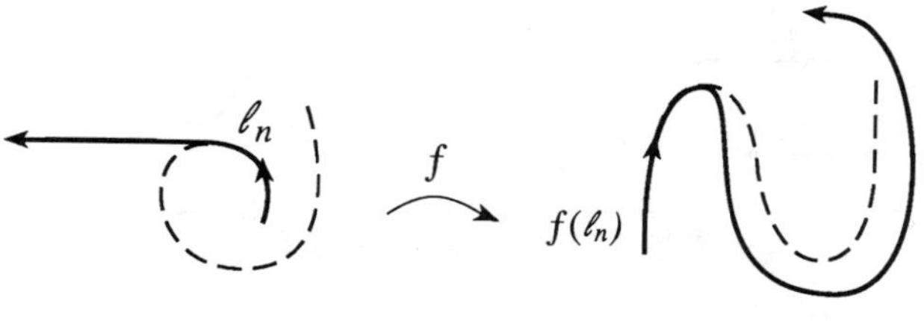

Figure 14.

In either case, points before the end point of ℓ_n are jump points. Hence rule (2) would apply before the point of horizontal tangency was reached. The case where ℓ_n is on a downward oriented right component of Γ is symmetric. □

Finally we must study the jump points, particularly the horizontal segments from a jump point and the character of the component of Γ on which it ends.

LEMMA 11. *Suppose n is odd so that ℓ_n is being formed by following Γ. Suppose z is a jump point of ℓ_n and the component of Γ in $\ell_n \cap \Gamma$ ending at z is either upward oriented left or downward oriented right in Γ_N. Let w be the point to which z jumps (as in rule (2)). Then*

1) *w is a local* min *of Γ if z is a left point or a local* max *of Γ if z is a right point,*
2) *the component of Γ which is connected to w in the direction of the tangent to w is upward oriented left if z is left, downward oriented right if z is right and is in Γ_N,*
3) *The segment $\overline{zw}$ is positive index (i.e. $\overline{zw} - f(\overline{zw})$ is a positive index loop).*

Proof. Condition (1) follows because ℓ_n does not begin at a jump point (see lemma 9), hence jump points arrive only when w is a point of horizontal tangency of Γ.

Condition (2) follows by the choice of orientation in step I and the fact that any rotation of the points in the same component of Γ as w and near w would result in jump points on ℓ_n before z.

Condition (3) follows exactly as lemma (5). □

Step IV: To complete the proof we note the following:

- •) for n odd, the portion of ℓ_n contained Γ are all in Γ_N (rule (3) and lemma 8, 11).
- •) for n odd, if ℓ_n is in left components of Γ then it is oriented upwards, if ℓ_n is n right components of Γ then it is oriented downwards (lemmas 7,8,10).
- •) $\ell_n - f(\ell_n)$ is always positive index (rule (3) and lemmas 5,6,7,8,10,11)
- •) if we construct the segments $\ell_0, \ldots \ell_n, \ldots$ as above, eventually a closed loop is formed (lemma (3) genericity assumption on Γ).

This closed loop is the required loop L which by lemma 2 completes the proof for maps satisfying the genericity assumptions of lemma 3. Again noting that the limit of maps on with a fixed point will have a fixed point, we obtain the desired result.

Concluding remarks. 1) The second theorem of Birkhoff actually yields two periodic orbits. Poincaré conjectures existence of two fixed points, but he only discusses the case of "generic" Γ. We do not know if there is a direct proof of existence of the second fixed point using these techniques.

2) It would also be interesting to know what Γ looks like for a generic twist map. Computer studies show it can be quite complicated, even for simple maps.

REFERENCES

[B1] G.D. Birkhoff, *Proof of Poincaré's last Geometric Theorem*, Trans. A.M.S., 14 (1913), pp. 14-22.

[B2] G.D. Birkhoff, *Sur la démonstration directe du dernier théorème géométrique de Henri Poincaré par M. Dantzig*, Bull. Sci. Math., 42 (1918), pp. 41-42.

[B3] G.D. Birkhoff, *An extension of Poincaré's last geometric theorem*, Acta Math., 47 (1925), pp. 297-311.

[BN] M. Brown and W.D. Neuman, *Proof of the Poincaré-Birkhoff fixed point theorem*, Michigan Math. J., 24 (1977), pp. 21-31.

[DeM-P] W. de Melo and J. Palis, *Geometric theory of dynamical systems*, Springer, 1982.

[F1] J. Franks, *Recurrence and fixed points of surface homeomorphisms*, Erg. Th. and Dyn. Sys., 8* (1988), pp. 99-108.

[F2] J. Franks, *A variation on the Poincaré-Birkhoff theorem*, in Hamiltonian Dynamical Systems, Eds. K. Meyer and D. Saari, Contemp. Math. A.M.S., 81 (1988), pp. 111-118.

[P] H. Poincaré, *Sur un théorème de géométrie*, Rend.del. Circ. Math. Palermo, 33 (1912), pp. 375-407.

DYNAMICS CONNECTED WITH INDEFINITE NORMAL TORSION

MICHAEL R. HERMAN*

Abstract. We prove that Lagrangian tori, invariant by exact symplectic C^∞ diffeomorphisms of the cotangent bundle of the n-torus that are perturbations of completely integrable monotone indefinite diffeomorphisms, have very different properties from the monotone positive case ([H3]). In particular Birkhoff's theorems and the uniqueness results (see [H5] and [H3]) do not generalize.

Studying many examples shows that the geometry of the Lagrangian invariant tori is very different from the one, one obtains for non integrable monotone twist maps of cotangent bundle of the circle.

Key words. Invariant Lagrangian tori, symplectic diffeomorphisms, completely integrable diffeomorphism, indefinite normal torsion.

AMS(MOS) subject classifications. 58F05, 58F07, 58F11.

1. Introduction

We want to study C^∞ n-tori, invariant by exact symplectic diffeomorphisms F of $T^*(\mathbf{T}^n) = \mathbf{A}^n$ that are C^∞ perturbations of a C^∞ completely integrable monotone indefinite diffeomorphism L, notion defined in 4.1 and 4.2. We show that properties of invariant Lagrangian tori are very different from the results obtained in [H_3] when L is monotone positive (definite).

In 3.2 to 3.5 we will define the notion of normal torsion (and its type) of F on an invariant Lagrangian torus T such that $F_{|T}$ is C^∞ conjugated to an ergodic translation. In 3.6, we recall the invariant tori theorem (with a slightly more general formulation).

In 4.6 we give examples where monotone indefinite completely integrable diffeomorphisms can occur. This can be the case for periodic elliptic points for perturbations of L, even when L is monotone positive definite.

In 4.7 to the end of paragraph 4 we illustrate the monotone indefinite phenomenon by special examples (non generic by 4.13). The examples of 4.7 (that generalizes the examples of [H_3,I.6]) are non completely integrable by 4.10, but have very different properties compared to non completely integrable monotone twist maps of $\mathbf{A}$. All the invariant tori can be calculated. In 4.14 we introduce the notion of lines of escape (justified by the Remark 4.19.1). The examples of 4.20 show that an open and dense set of $\mathbf{A}^n$ can be the set of wandering points of an exact symplectic diffeomorphism F.

The theorem 5.2 shows that Birkhoff's theorems (see [H_3,I.4] and [H_5]) cannot be generalized to C^∞ perturbations of any L that is monotone indefinite. In 5.18 we study the very similar case $n = 1$, when degenerate torsion occurs.

The theorem 6.4 shows that the type of the normal torsion, when indefinite, is very instable by C^∞ perturbations and it can radically change. As a consequence

*Centre de Mathematiques, Unité de Recherche Associée au CNRS n° 169, Ecole Polytechnique, F-91128 Palaiseau Cedex (FRANCE).

in 6.29, we obtain a non uniqueness result of invariant tori with a fixed rotation vector in the universal covering of $\mathbf{A}^n$. This differs from the results of [H$_3$, I.10].

In 7, we introduce a subset $IT^\infty(F)$ of the C^∞ Lagrangian tori τ, homotopic to $\{r = 0\}$, invariant by F and such that $F_{|\tau}$ is C^∞ conjugated to a diophantine translation. We only consider that the tori τ included in the compact set $\mathbf{A}^n_{\delta_1}$ for a fixed $\delta_1 > 0$. Then we take limits in the Hausdorff topology. We state various results (along the lines of [H$_6$]) and show that 5 and 6 yield generic properties and 7.11 give that the generic dynamics results stated in [H$_6$] also hold.

We never define monotone indefinite diffeomorphisms of $\mathbf{A}^n$. This contrasts with [H$_3$,I] where monotone positive (definite) diffeomorphisms where defined. Both 6 and 7 show what ever definition is given, it would not be very useful. We define the indefinite normal torsion of an invariant Lagrangian torus on which the diffeomorphism is C^∞ conjugated to a diophantine translation and study C^∞ perturbations of the diffeomorphism. This is the theme of this work.

I would like to thank Jean-Christophe Yoccoz for many fruitful discussions, the University of Minnesota and the I.M.A. for its invitation and support for a very pleasant visit during which this work was presented, Patrice Le Calvez for helping me improve the text and Madame Paule Truc for typing, with great care, the manuscript.

2. Notations and definitions

2.1. Most notations are the same as for [H$_3$] and [H$_4$].

When $n \in \mathbf{N}^*$ we denote by $\mathbf{T}^n = \mathbf{R}^n/\mathbf{Z}^n$ the torus of dimension n, by $R_\alpha : \theta \to \theta + \alpha$ the translations of $\mathbf{T}^n$ or $\mathbf{R}^n$, and by $d\theta = d\theta_1 \otimes \cdots \otimes d\theta_n \equiv m$ the Haar measure of $\mathbf{T}^n$.

The cotangent bundle of $\mathbf{T}^n$ is noted by $T^*(\mathbf{T}^n) \cong \mathbf{T}^n \times \mathbf{R}^n \cong \mathbf{A}^n$ where $\mathbf{A} = T^*(\mathbf{T}^1)$. The action angle coordinates on $\mathbf{A}^n$ are written $(\theta, r), \theta = (\theta_1, \ldots, \theta_n)$, $r = (r_1, \ldots, r_n)$. The Liouville 1 form is the 1 form $v_n = \sum_1^n r_j d\theta_j$ and $w_n = -dv_n$ is the canonical symplectic form on $\mathbf{A}^n$.

A C^∞ diffeomorphism F of $\mathbf{A}^n$ is called *symplectic* if $F^* w_n = w_n$, and *exact symplectic* when the 1 form $F^* v_n - v_n$ is exact.

2.2. When F is a C^∞ diffeomorphism of $\mathbf{A}^n$, homotopic to the identity, the following statements are equivalent :

- The diffeomorphism F is exact symplectic.
- The 1 form $F^* v - v$ is exact where v is any C^∞ 1 form such that $dv = -w_n$.
- The diffeomorphism $H \circ F \circ H^{-1}$ is exact symplectic for any C^∞ symplectic diffeomorphism H of $\mathbf{A}^n$.

In the following f^k, $k \in \mathbf{Z}$ denotes the kth iterate of any homeomorphism of topological space X. A set X_1, $X_1 \subset X$ is invariant by f if $f(X_1) = X_1$.

2.3. If F is a C^∞ diffeomorphism of $\mathbf{A}^n$, we write its Jacobian matrix in the coordinates

$$x = (\theta, r) \quad , \quad DF(x) = \begin{pmatrix} a(x) & b(x) \\ c(x) & d(x) \end{pmatrix}$$

where a, b, c, d are C^∞ mappings of $\mathbf{A}^n$ into $M_n(\mathbf{R})$ = the $n \times n$ matrices with coefficients in $\mathbf{R}$. We also denote by Df the Jacobian matrix of a diffeomorphism of $\mathbf{T}^n$.

We use the notation C^ω for $\mathbf{R}$-analytic, $\frac{\partial}{\partial\theta} = \left(\frac{\partial}{\partial\theta_1}, \cdots, \frac{\partial}{\partial\theta_n}\right)$ and $\frac{\partial}{\partial r} = \left(\frac{\partial}{\partial r_1}, \cdots, \frac{\partial}{\partial r_n}\right)$.

2.4. In the following a C^k torus of $\mathbf{A}^n$ is the image by a C^k embedding of $\mathbf{T}^n$ into $\mathbf{A}^n$. We will only consider, in all this article, tori of dimension n.

If $\psi \in C^0(\mathbf{T}^n, \mathbf{R}^n)$ the *graph* of ψ is noted $\Gamma_\psi = \{(\theta, \psi(\theta)), \theta \in \mathbf{T}^n\}$.

On $\mathbf{R}^n$ we put the Euclidian norm $\|(x_j)\| = (\sum_1^n x_j^2)^{1/2}$ and on $M_n(\mathbf{R})$ the induced operator norm $\|a\| = \sup_{\|v\| \leq 1} \|av\|$, $v \in \mathbf{R}^n$ and $a \in M_n(\mathbf{R})$.

We identify $a \in M_n(\mathbf{R})$ with the associated linear map $v \in \mathbf{R}^n \to av \in \mathbf{R}^n$. Of course this identification supposes that $v = \begin{pmatrix} v_1 \\ \vdots \\ v_n \end{pmatrix} \in \mathbf{R}^n$ is a column vector but for obvious typographical reasons we will write $v = (v_1, \ldots, v_n)$.

In the following $\|\varphi\|_{C^0} = \sup_{x \in X} \|\varphi(x)\|$ where $\varphi : X \to \mathbf{R}^n$ or $M_n(\mathbf{R})$ is a function.

If Y is a subset of a topological space we denote by $\mathrm{Int}(Y)$ its interior and by $\overline{Y}$ its closure.

3. The existence of invariant tori (KAM)

3.1. The vector $\alpha = (\alpha_1, \ldots, \alpha_n) \in \mathbf{T}^n$ satisfies a *diophantine condition*, and we write $\alpha \in DC$, if there exists $\beta > 0$ and $\gamma > 0$ such that for every $k = (k_1, \ldots, k_n) \in \mathbf{Z}^n - \{0\}$ we have $|||\langle k, \alpha\rangle||| \geq \gamma \|k\|^{-n-\beta}$ where $\langle k, \alpha\rangle = \sum k_j\alpha_j$ and $|||\ |||$ denotes the distance to the nearest integer : if $\tilde{x} \in \mathbf{R}$ projects to $x \in \mathbf{T}^1$, $|||x||| = \inf_{p \in \mathbf{Z}} |\tilde{x} + p|$.

We say that the translation R_α satisfies a diophantine condition if $\alpha \in DC$.

We have $\mathbf{T}^n = DC \amalg \mathcal{R} \amalg \mathcal{L}$ where $\mathcal{R} = \{\alpha \in \mathbf{T}^n ; \exists\, k \in \mathbf{Z}^n - \{0\}, \langle k, \alpha\rangle = 0\}$. The set $\mathcal{L}$ is by definition the set of Liouville vectors in $\mathbf{T}^n$. The set $\mathcal{R} \cup \mathcal{L}$ has Haar measure 0 but $\mathcal{L}$ is a dense G_δ of $\mathbf{T}^n$ (see [H$_4$,V]).

A G_δ set is, by definition, countable intersection of open sets.

3.2. If $\delta' > 0$, we define $\mathbf{A}^n_{\delta'} = \{(\theta, r) \in \mathbf{A}^n ; \|r\| \leq \delta'\}$.

Let $L : \mathbf{A}^n_{\delta'} \to \mathbf{A}^n$ be a C^∞ symplectic embedding leaving invariant $\mathbf{T}^n_0 \equiv \mathbf{T}^n \times \{0\} \equiv \{r = 0\}$ and we suppose that $L_{|\{r=0\}} = R_\alpha$ where R_α is an ergodic translation of $\mathbf{T}^n$.

We have

$$L(\theta, r) = (\theta + \alpha + b(\theta)r + O(r^2), r + O(r^2))$$

when $r \to 0$. The function $\theta \to b(\theta) \in M_n(\mathbf{R})$ is C^∞ and ${}^t b = b$, where ${}^t b$ denotes the transpose matrix at every point. This follows from the fact that L is symplectic (see [H_2,I.2]).

Let $b_1 \in M_n(\mathbf{R})$, $b_1 = \int_{\mathbf{T}^n} b(\theta)d\theta$.

We say that L has *non degenerate torsion* on $\{r = 0\}$ if $\det(b_1) \neq 0$ where $\det(b_1)$ denotes the determinant of the matrix b_1. Under the above hypothesis, the symmetric matrix b_1 has a type[1] : it can be definite or indefinite when $n \geq 2$.

3.3. We can reduce to the above case when we replace $L_{|\{r=0\}} = R_\alpha$ by $L_{|\{r=0\}} = f = h \circ R_\alpha \circ h^{-1}$ where h is a C^∞ diffeomorphism of $\mathbf{T}^n$. For this, we conjugate F by the symplectic diffeomorphism (see 3.10)

$$H(\theta, r) \equiv T^* h(\theta, r) \equiv (h(\theta), {}^t(Dh(\theta))^{-1} r)$$

and consider

$$H^{-1} \circ L \circ H(\theta, r) = (\theta + \alpha + b(\theta)r + O(r^2), r + O(r^2)), \qquad \text{when} \quad r \to 0\,.$$

The matrix b_1 does not depend on the choice of h (by hypothesis the translation R_α is ergodic) : We obtain the same matrix b_1 if we conjugate $H^{-1} \circ L \circ H$ by a C^∞ symplectic diffeomorphism G leaving invariant $\{r = 0\}$ and satisfying $G_{|\{r=0\}} = R_\beta$, $\beta \in \mathbf{T}^n$.

Remark. The fact that we suppose that the translation R_α is ergodic is necessary to define the normal torsion as we will see in 6.28.1.

Furthermore if the normal torsion is degenerate higher order terms in the Birkhoff normal form (see 3.8) are necessary to study invariant tori. The normal torsion is the first term of the Birkhoff normal form.

3.4. More generally, if F is a C^∞ symplectic diffeomorphism of $\mathbf{A}^n$ that leaves invariant a C^∞ n-torus $T \subset \mathbf{A}^n$ such that $F_{|T}$ is C^∞ conjugated to an ergodic translation R_α of $\mathbf{T}^n$, then by [H_2,I.3] the torus T is Lagrangian.
Then by a theorem of A. Weinstein [W_1], there exists a neighbourhood of T in $\mathbf{A}^n$ that is symplectically C^∞ isomorphic to a neighbourhood of the 0 section of $T^*(\mathbf{T}^n)$ the isomorphism taking T to $\{r = 0\}$. We can then proceed as in 3.2 and 3.3 and define a matrix b_1. The fact that $\det(b_1) \neq 0$ or the type of b_1 does not depend on the possible choices (cf. 3.2). We therefore say that F has non degenerate torsion on T, if $\det(b_1) \neq 0$, and then we can define the type of the torsion of F on T.

Remark. Given a C^∞ n-torus T and a C^∞ diffeomorphism f of T, then f is C^∞ conjugated to a translation if there exists a C^∞ diffeomorphism $h : \mathbf{T}^n \to T$ such that $h^{-1} \circ f \circ h = R_\alpha$. The vector $\alpha \in \mathbf{T}^n$ is not well defined since if $A \in GL(n, \mathbf{Z})$, $h \circ A^{-1}$ gives the rotation $R_{A\alpha}$. Conjugating by $T^* A^{-1}$, when the translation R_α is ergodic, yields the normal torsion $A b_1 {}^t A$. If one fixes the homotopy class of

[1] In the following we will associate to a symmetric matrix b_1 the quadratic form $v \in \mathbf{R}^n \to {}^t v b_1 v$. Definite and indefinite always implies, by definition, non degenerate.

the conjugacy $h : \mathbf{T}^n \to T$ then the vector α and the normal torsion b_1 are well determined. Unfortunately there is no canonical way of choosing a homotopy : it is like the rotation number of an orientation preserving homeomorphism of a circle, it requires, to be defined, the choice of an orientation on the circle.

In the following all the n-tori in $\mathbf{A}^n$ will be homotopic to $\{r = 0\}$ and as we have, since the beginning, taken $\mathbf{T}^n$ that has a canonical basis, we can therefore choose canonically a homotopy class for h : the one such that h is homotopic to the canonical inclusion $\mathbf{T}^n \hookrightarrow \mathbf{T}^n \times \{0\} \subset \mathbf{A}^n$. With this choice we can determine the rotation R_α and the normal torsion.

3.5. When T is the graph of $\psi \in C^\infty(\mathbf{T}^n, \mathbf{R}^n)$, as the torus T is Lagrangian, the C^∞ diffeomorphism $K(\theta, r) = (\theta, r + \psi(\theta))$ is symplectic. Considering $K^{-1} \circ F \circ K$ reduces the calculation to the case 3.3. The reader should notice that 2.2 applies and $K^{-1} \circ F \circ K$ is exact symplectic.

3.6. The invariant tori theorem

We suppose that L satisfies the hypothesis of 3.2 and furthermore that $\alpha \in DC$ and L has a non degenerate normal torsion on $\{r = 0\}$ (i.e. $\det(b_1) \neq 0$).

THEOREM. *There exists a neighbourhood V of L in $C^\infty(\mathbf{A}^n_{\delta'}, \mathbf{A}^n)$, for the C^∞ topology, such that if $F \in V$ and F is a C^∞ exact symplectic embedding of $\mathbf{A}^n_{\delta'}$ into $\mathbf{A}^n$ then F leaves invariant the torus $\Gamma_{\psi_F} = T_F$ where $\psi_F \in C^\infty(\mathbf{T}^n, \mathbf{R}^n)$ and $F_{|T_F}$ is C^∞ conjugated to R_α by a C^∞ diffeomorphism $h_F : \mathbf{T}^n \to T_F$. Moreover the map $F \in V \to (\psi_F, h_F)$ is continuous for the C^∞ topology and $(\psi_L, h_L) = (0, \mathrm{Id}_{\mathbf{T}^n})$.*

Remark. To determine h_F we use 3.4 and not only is the map $F \to (\psi_F, h_F)$ continuous for the C^∞ topology (on all spaces), but it is also a C^∞ tame map as defined by R. Hamilton (cf. [B$_2$]). When h_F is suitably normalized, this map is a local C^∞ diffeomorphism from a neighbourhood of L onto a neighbourhood of $(0, \mathrm{Id}_{\mathbf{T}^n})$. This implies a local uniqueness result : the one given by an implicit function theorem.

3.7. The reader will find in the survey article of J-B. Bost B$_3$] many references (that we will not reproduce here) and a proof of the invariant tori theorem for Hamiltonian vector fields (KAM). The case of diffeomorphisms (the above theorem of 3.6) can be reduced [D$_2$] to the case of Hamiltonian vector fields, when L is completely integrable (i.e. $L(\theta, r) = (\theta + \ell(r), r)$) and satisfies $\ell(0) = \alpha$ and det $D\ell(0) \neq 0$. One can then deduce 3.6 from Birkhoff's normal form (3.8). It turns out that one can directly prove 3.6 by the methods of [B_3].

The formulation of 3.6 is more global than the one usually stated. We can apply 3.4 and an torus T, invariant by F, can be continued (for exact symplectic C^∞ perturbations) as long as T is C^∞, $F_{|T}$ is C^∞ conjugated to R_α and the normal torsion of F on T is non degenerate. For the fact that invariant tori can break down, the reader can consult [H$_8$] and [H$_5$]. In the examples of [H$_5$] and [H$_8$] the normal torsion stays definite and the break down is associated to a loss of smoothness. When $n \geq 2$, in the indefinite case, the normal torsion can become degenerate, see [6.4].

3.8. Birkhoff's normal form

We consider a symplectic embedding L satisfying the hypothesis of 3.2. We suppose furthermore $\alpha \in DC$. The reader will find in [D$_1$][D$_2$][B$_2$] the proof of the following proposition.

PROPOSITION. *For every integer $k \geq 1$ one can construct a symplectic embedding G_k of a neighbourhood of $\{r = 0\}$ in $\mathbf{A}^n_{\delta'}$, $G_k(\theta, 0) = (\theta, 0)$, such that one has*

$$G_k^{-1} \circ L \circ G_k(\theta, r) = (\theta + \alpha + \ell_k(r) + O(r^{k+1}),\, r + O(r^{k+1}))$$

when $r \to 0$, where $\ell_k(r) = b_1 r + O(r^2)$ is a polynomial map of degree k depending only on the variables $r_1, \dots, r_n$ and $L_k : (\theta, r) \to (\theta + \alpha + \ell_k(r), r)$ is a symplectic diffeomorphism.

3.9. We suppose that the hypothesis of 3.8 are satisfied.

PROPOSITION. *There exists a sequence of C^∞ exact symplectic diffeomorphisms $(H_j)_{j\in\mathbf{N}}$ of $\mathbf{A}^n$, $H_j \to \mathrm{Id}_{\mathbf{A}^n}$ in the C^∞ topology, such that each H_j leaves invariant $\{r = 0\}$, $H_{j|\{r=0\}} = \mathrm{Id}_{\mathbf{T}^n}$ and $L \circ H_j$ has a non degenerate normal torsion on $\{r = 0\}$.*

This follows from the proposition of 3.8 (with $k = 1$) by considering $G_1^{-1} \circ L \circ G_1 \circ \widehat{L}_j$ where $\widehat{L}_j$ are C^∞ symplectic diffeomorphisms of the form $(\theta, r) \to (\theta + \ell_j(r), r)$, $\widehat{L}_j \to \mathrm{Id}$ as $j \to +\infty$ and satisfying, $\ell_j(0) = 0$, $\det(D\ell_j(0) + b_1) \neq 0$ and $\ell_j(r) = 0$, if $\|r\| \geq \varepsilon > 0$, where $\varepsilon > 0$ is chosen small enough so that $H_j = G_1 \circ \widehat{L}_j \circ G_1^{-1}$ is defined and extends by the identity of $\mathbf{A}^n$ outside a small neighbourhood of $\{r = 0\}$.

3.10. Let f be a C^∞ diffeomorphism of $\mathbf{T}^n$, then the cotangent map of f

$$T^* f(\theta, r) = (f(\theta),\, {}^t(Df(\theta))^{-1} r)$$

is symplectic since it preserves the Liouville v_n one form defined in 2.1. It leaves invariant the torus $\{r = 0\} \equiv \mathbf{T}^n \times \{0\} \equiv \mathbf{T}^n_0$ and $T^* f_{|\{r=0\}} = f$.

PROPOSITION. *Given $\varepsilon > 0$ and f a C^∞ diffeomorphism of $\mathbf{T}^n$, C^1 isotopic to the identity, then there exists F_ε a C^∞ exact symplectic diffeomorphism of $\mathbf{A}^n$ such that*

$$F_\varepsilon = T^* f \quad \text{in a neighbourhood of } \{r = 0\}$$

and

$$F_\varepsilon(\theta, r) = (\theta, r), \text{ if } \|r\| \geq \varepsilon.$$

Furthermore F_ε tend to Id_{A^n} in the C^∞ topology if f tends to $\mathrm{Id}_{\mathbf{T}^n}$ in the C^∞ topology.

The proof is folklore using generating functions. When f is C^1 close to the identity one can define, as in [D$_2$, p.34-35] and with the same notations, a generating function $S(\theta, R)$ near $\{r = 0\}$, such that $T^* f(\theta, r) = (\Theta, R)$ is given by the following formulas (valid near $\{r = 0\}$) :

$$\begin{cases} \Theta_i = \theta_i + \frac{\partial S}{\partial R_i}(\theta, R) \\ R_i = r_i - \frac{\partial S}{\partial \theta_i}(\theta, R) \end{cases} \qquad i = 1, 2 \ldots, n.$$

One defines an extension F_ε, by using the same formulas replacing the generating function S by the function $\sigma(R)S(\theta, R)$ where $\sigma : \mathbf{R}^n \to \mathbf{R}$ is a C^∞ function equal to 1, in a neighbourhood of 0, and equal to 0, if $\|r\| \geq \varepsilon'$, $0 < \varepsilon' \ll \varepsilon$. The continuity statement follows. When f is C^1 isotopic to the identity, f is a finite composition of diffeomorphisms C^1 (or even C^∞) close to the identity. The extension of f is obtained by finite composition of the extensions we constructed above.

4. Monotone indefinite completely integrable diffeomorphisms: The definition and examples of perturbations

4.1. In the following, L will denote a C^∞ *completely integrable* diffeomorphism of $\mathbf{A}^n$: L is symplectic and of the form $L(\theta, r) = (\theta + \ell(r), r)$. The fact that L is symplectic is equivalent to $\ell(r) = \frac{\partial \hat{\ell}(r)}{\partial r}$ where $\hat{\ell} \in C^\infty(\mathbf{R}^n, \mathbf{R})$ and $\frac{\partial}{\partial r} = (\frac{\partial}{\partial r_1}, \ldots, \frac{\partial}{\partial r_n})$. It follows that the matrix $D\ell(r)$ is symmetric (i.e. ${}^tD\ell(r) = D\ell(r)$) for every $r \in \mathbf{R}^n$. The matrix $D\ell(r)$ is called the (normal) torsion of L. We also write $L_{\hat{\ell}} = L$ to indicate the dependance on $\hat{\ell}$.

We say L is *monotone* (resp. monotone on $\mathbf{A}^n_\delta$) if $\det(D\ell(r)) \neq 0$ for every $r \in \mathbf{R}^n$ (resp. for $\|r\| \leq \delta$). We choose the word monotone instead of the longer expression non degenerate torsion. If L is monotone and $a \in \mathbf{R}^n$ satisfies $\ell(a) = \alpha \in DC$ then the torus $\{r = a\}$ can be continued for C^∞ exact symplectic perturbations of L (i.e. 3.6 applies after conjugating by $(\theta, r) \to (\theta, r + a)$).

Example. The diffeomorphism $(\theta, r) \to (\theta + Br, r)$ of $\mathbf{A}^n$ with $B \in M_n(\mathbf{R})$ is symplectic if B is symmetric and is monotone if $\det(B) \neq 0$.

4.2. Given L monotone on $\mathbf{A}^n_\delta$ the quadratic form $Q_L(r) : v \in \mathbf{R}^n \to {}^t v D\ell(r) v = D^2\hat{\ell}(r)(v, v)$ is non degenerate on $\|r\| < \delta$ and hence has a type. It is *definite* if $n = 1$, but when $n \geq 2$, it can be *indefinite* and we say that L is *monotone indefinite* on $\mathbf{A}^n_\delta$. In the last case, for any r, $\|r\| < \delta$, we can find an *isotropic vector* $v \neq 0$ for $Q_L(r)$, that is $Q_L(r)(v) = 0$.

Example. Take $n = 2$ and $L_B(\theta, r) = (\theta + Br, r)$ with $B = \begin{pmatrix} 0 & 1 \\ 1 & 0 \end{pmatrix}$, then L_B is monotone indefinite.

We can generalize the above example in the following way : let $\hat{\ell} \in C^\infty(\mathbf{R}^n, \mathbf{R})$ of the form

$$\hat{\ell}(r_1, \ldots, r_n) = r_1 \ell_1(r_2, \ldots, r_n) + \hat{\ell}_2(r_2, \ldots, r_n) \tag{4.3'}$$

and such that $\det D\ell(r) \neq 0$ for every $r \in \mathbf{R}^n$ (e.g. $\hat{\ell}$ can be a non degenerate quadratic form). The completely integrable diffeomorphism

$$L(\theta, r) = (\theta_1 + \ell_1(r_2, \ldots, r_n)\,,\, \theta_2 + \frac{\partial \hat{\ell}}{\partial r_2}(r), \ldots, r) \tag{4.3}$$

is monotone indefinite.

4.4. We can reformulate the definition in the following more geometric way :

Let $L_{\hat{\ell}}$ be monotone, then $L_{\hat{\ell}}$ is monotone indefinite iff the hypersurface $\Gamma_{\hat{\ell}} = \{(r, \hat{\ell}(r)) \in \mathbf{R}^n\,;\, r \in \mathbf{R}\}$ of $\mathbf{R}^{n+1}$, with the induced Riemanniann metric from the Euclidian metric $\|\ \ \|$ on $\mathbf{R}^{n+1}$, has at each point a strictly negative part in its sectional curvature tensor (i.e. the function $\hat{\ell}$ is not strictly convex, or concave) :

At every point $x \in \Gamma_{\hat{\ell}}$, $x = (r, \hat{\ell}(r))$ we can find a tangent vector $v_1 = (v, \langle \frac{\partial \hat{\ell}}{\partial r}(r), v\rangle)$ $v \in \mathbf{R}^n - \{0\}$ that defines an asymptotic direction of $\Gamma_{\hat{\ell}}$ at x (it is enough to take an isotropic vector of $Q_{L_{\hat{\ell}}}(r)$). For the definition of an asymptotic direction we refer to [H_9].

4.5. Hamiltonian flows. We start with $H \in C^\infty(\mathbf{R}^n, \mathbf{R})$ and consider the Hamiltonian $H(r)$ on $\mathbf{A}^n$,then we obtain the completely integrable Hamiltonian flow associated to H by integrating the vector field $X_H = (\frac{\partial H}{\partial r}, -\frac{\partial H}{\partial \theta})$:

$$F_s : (\theta, r) \to \left(\theta + s\frac{\partial H}{\partial r}(r), r\right), \quad s \in \mathbf{R}.$$

If we want to perturb the Hamiltonian $H(r)$ and study C^∞ invariant Lagrangian tori, as these tori have to lie in a surface of energy, the notion of indefinite torsion should use this, especially when $n = 2$.

We only look at neighbourhood of $r = 0$. We can always suppose that $H(0) = 0$. There are 2 cases :

(1) • $\frac{\partial H}{\partial r}(0) = 0$. This case is singular and we ask that $F_{|s=1}$ is indefinite near $\{r = 0\}$. This can happen even when $n = 2$. The examples of 4.9.2 (extracted from [H_2]) and the proof of 5.2 will convince the reader that the indefinite phenomenon can occur.

(2) • $\frac{\partial H}{\partial r}(0) = (\alpha_1, \ldots, \alpha_n) \neq 0$. Then we look at a Poincaré section. Let us suppose $\alpha_n \neq 0$ and take the Poincaré section $S = \{(\theta, r); \theta_n = 0, H(r) = 0, \|r\| \text{ small }\}$. We obtain as 1st return map the C^∞ completely integrable diffeomorphism $L : \mathbf{A}_\delta^{n-1} \to \mathbf{A}_\delta^{n-1}$ for some $\delta > 0$.

We have $L(\theta, r) = (\theta + \frac{\partial \hat{\ell}}{\partial r}(r), r)$, where the function $\hat{\ell}$ is given, using the implicit function theorem, by

$$H(r_1, \ldots, r_{n-1}\,,\, \hat{\ell}(r_1, \ldots, r_{n-1})) = 0 \quad \text{and} \quad \hat{\ell}(0) = 0.$$

The conjugation is obtained by the embedding

$$(\theta, r) \in \mathbf{A}_\delta^{n-1} \to (\theta, 0, r, \hat{\ell}(r)) \subset S\,,$$

the pull back of the 2 form w_n being w_{n-1} where w_n is defined in 2.1. For the study of Hamiltonians that are perturbations of $\hat{\ell}$ near $\{0\}$, we refer the reader to [D_2,I].

The type of the torsion of L at $r = 0$ depends only on the geometry of the hypersurface $\{r \in \mathbf{R}^n; H(0) = 0\}$ near $r = 0$, it is monotone iff the Gaussian curvature of $H^{-1}(0)$ at 0 is non zero and under the above hypothesis it is definite (resp. indefinite) if $H^{-1}(0)$ is the graph of a convex function near 0 (resp. i.e. it is not the graph of a convex function near 0). When $n = 2$, it is definite or degenerate. We leave the proofs to the reader. Keeping in mind what we just said, we will show elsewhere how to adapt everything we will do to Hamiltonian flows.

4.6. Examples where indefinite torsion can occur

1. We identify $T^*(\mathbf{R}^n) = \mathbf{R}^n \oplus (\mathbf{R}^n)^*$ with $\mathbf{C}^n$ and consider the Hamiltonian $H = \sum_1^n a_j|z_j|^2 + \sum_1^n b_j|z_j|^4$, $a_j \in \mathbf{R}^*$ and $b_j \in \mathbf{R}$, $1 \leq j \leq n$, where $z = (z_1, \dots, z_n) \in \mathbf{C}^n$. In symplectic polar coordinates near 0, that is $z_j = \frac{1}{\pi}\sqrt{r_j}\, e^{2\pi i\theta_j}$, $1 \leq j \leq n$, the Hamiltonian H transforms to the Hamiltonian $\hat{H}$ on $T^*(\mathbf{T}^n)$

$$\widehat{H}(r) = \frac{1}{\pi}\sum_1^n a_j r_j + \frac{1}{\pi^2}\sum_1^n b_j r_j^2\,, \quad r_j > 0$$

and the torsion of time one map of the Hamiltonian flow associated to $\widehat{H}$ is equal to $\frac{2}{\pi^2}$ diag$(b_1, \dots, b_n)$. It does not depend on the numbers a_j, $1 \leq j \leq n$.

If $a_j > 0$, $1 \leq j \leq n$, near 0, the function H is convex.

We leave the reader to do the computations of 4.5, when $n \geq 3$, and check that there is no simple relation between the signs of the numbers a_j and the type of the first return map of the flow associated to $\hat{H}$ of a Poincaré section.

If one considers a C^∞ Hamiltonian on $\mathbf{C}^n$ near a local generic[2] minimum $0 \in \mathbf{C}^n$, the type of the torsion[3] of the Birkhoff normal form in symplectic polar coordinates depends on the derivatives up to order 4 of H at 0 and the author of these lines has no geometric intuition how one can guess the type of the torsion from geometric properties of H.

2. If one considers C^∞ symplectic perturbations of the completely integrable diffeomorphism of $\mathbf{A}^n$, $L : (\theta, r) \to (\theta + r, r)$ that is monotone positive definite, one might ask what happens to the type of the periodic elliptic (that we suppose generic) after the Birkhoff normal form. The following example shows that they can be indefinite, when $n \geq 2$.

Example. Let $H = \frac{\|r\|^2}{2} + \varepsilon V(\theta)$, $\varepsilon > 0$, be a C^∞ Hamiltonian on $T^*(\mathbf{T}^n)$. We suppose that the C^∞ function V is equal to $\sum_1^n a_j\theta_j^2$, $a_j > 0$, in a neighbourhood of 0. Considering the time one map F of the Hamiltonian flow associated to H, we can perturb F in the C^∞ topology, using the examples 1 above, such that near the elliptic fixed point 0 of F, after the Birkhoff normal form and in symplectic polar coordinates, F has any type giving in advance.

The above examples, showing that generic elliptic fixed points can be indefinite after the Birkhoff normal form and in symplectic polar coordinates, for C^∞ perturbations of L can be generalized to any C^∞, completely integrable, monotone, diffeomorphism $L_{\hat{\ell}}$ such that $\frac{\partial \hat{\ell}}{\partial r}(0) = 0$. It is enough to consider Hamiltonians $H = \hat{\ell}(r) + \varepsilon V(\theta)$, $\varepsilon > 0$, $\varepsilon \to 0$. To see this, we remark after a linear symplectic change of coordinates (taking $\varepsilon = 0$) $\hat{\ell}(r) = \sum_1^n \varepsilon_j r_j^2 + O(r^3)$, when $r \to 0$, where

[2] The eigenvalues satisfy no non trivial resonance relation.

[3] We truncate symplectically the Birkhoff normal to order 4 and in symplectic polar coordinates, we suppose that torsion is non degenerate. This is a generic condition.

$\varepsilon_j = \pm 1$. In these coordinates we suppose that

$$V(\theta) = \sum_1^n \varepsilon_j a_j \theta_j^2 + \sum_1^n b_j \theta_j^4 + O(\theta^5),$$

when $\theta \to 0$, where $a_j > 0$, for $j = 1, \ldots, n$, are suitably chosen, $b_1 > 0$ and $b_j < 0$, if $2 \leq j \leq n$.

One then performs the Birkhoff normal form, after the linear change of variables $(\theta_j, r_j) = (\lambda_j \theta_j, \frac{1}{\lambda_j} r_j)$ with $\lambda_j = (\varepsilon a_j)^{1/4}$ and dividing the Hamiltonian by $\varepsilon^{1/2}$ and letting $\varepsilon \to 0$, the type of the torsion , up to sign, is the same as the diagonal matrix $\mathrm{diag}(b_1, \ldots, b_n)$.

The examples can be made C^ω by remarking that the Birkhoff normal form only depends on the 4th jet at one point.

3. The theory of relativity has lead mathematicians to study the geodesic flows on a C^∞ connected metrizable, manifold M when one replaces C^∞ Riemannian metric tensors by C^∞ semi (or pseudo) Riemanniann (non degenerate) metric tensors of signature (p, q). When $p.q \neq 0$, Riemannian metric tensors of signature (p, q) on M exists iff the tangent bundle of M has a reduction of its structural group to the group $\mathbf{O}(p, q, \mathbf{R})$. This is purely a question of algebraic topology and the answer is always positive if the tangent bundle of M is trivial. In the case of the tangent bundle of $\mathbf{T}^n$ this reduces, after a Legendre transform, to the study of C^∞ Hamiltonians H on $T^*(\mathbf{T}^n)$ such that for every θ, $r \to H(\theta, r)$ is a non degenerate quadratic form.

One nice fact is that these flows are not always complete (see [B$_1$]). We have the following example due to P.M. Williams.

Example. On $T^*(\mathbf{T}^2)$ we consider the Hamiltonian $H = r_1 r_2 + (\varepsilon \sin 2\pi\theta_1) r_1^2$, and one integrates the flow taking $r_2 = 0$ and $\theta_1 = 0$.

4.7. We consider L of the form 4.3 and $\varphi(\theta_1, \ldots, \theta_n) = \varphi(\theta_1)$ any function of $C^\infty(\mathbf{T}^1, \mathbf{R})$ such that an infinite number of its Fourier coefficients are non zero (e.g. $\varphi(\theta_1) = \varepsilon \sin(\sin 2\pi\theta_1))$, $\varepsilon \neq 0$). Let $G_\varphi(\theta, r) = (\theta, r + \frac{\partial \varphi}{\partial \theta}(\theta))$, $\frac{\partial}{\partial \theta} = \left(\frac{\partial}{\partial \theta_1}, \ldots, \frac{\partial}{\partial \theta_n}\right)$ and $F_\varphi = G_\varphi \circ L$. We studied in [H$_3$,I.6.2] the case $n = 2$ with $\ell(r) = Br$ and $B = \begin{pmatrix} 0 & 1 \\ 1 & 0 \end{pmatrix}$.

We look for invariant tori $\Gamma_{\hat{\psi}_a}$ by F_φ that are graphs of $\hat{\psi}_a(\theta_1, \ldots, \theta_n) = (\psi_a(\theta_1), a)$ with $\psi_\alpha \in C^\infty(\mathbf{T}^1, \mathbf{R})$ and $a \in \mathbf{R}^{n-1}$: ψ_a is given by

$$\psi_a \circ R_{\ell_1(a)} - \psi_a = \varphi' \circ R_{\ell_1(a)}\,, \quad \ell_1(a) \in \mathbf{R}. \tag{4.8}$$

This is possible when $\ell_1(a) \in DC$. Since L is monotone, the map $a \in \mathbf{R}^{n-1} \to \ell_1(a) \in \mathbf{R}$ is non constant on a neighbourhood of any point (this will also be true is $L \in C^\omega$ and the function ℓ_1 is non constant).

4.9. It follows from [H$_4$,XIII,4.11] that there exists a dense G_δ set $G_1 \subset \mathbf{R}^{n-1}$ such that when $a \in G_1$, $\ell_1(a) \notin \mathbf{Q}$ and 4.8 has no Lebesgue measurable solution. By a theorem of Gottschalk and Hedlund every orbit of F_φ on

$M_a = \{(\theta, r); (r_2, \ldots, r_n) = a\}$ is an unbounded. To see this, we remark that if $p_a : (\theta, r) \in M_a \to (\theta_1, r_1)$ we have $p_a \circ F_{\varphi|M_a} = f_a \circ p_a$ where

$$f_a : (\theta_1, r_1) \in \mathbf{A} \to (\theta_1 + \ell_1(a)), r_1 + \varphi'(\theta_1 + \ell_1(a)) \in \mathbf{A}.$$

This reduces to show that f_a has no bounded orbit when $a \in G_1$ (see [G$_1$] or [H$_4$,IV.4]). Furthermore f_a has a dense orbit on $\mathbf{A}$ (see [G$_1$]). We obtain therefore the properties :

(4.10) Almost every $x \in \mathbf{A}^n$ belongs to C^∞ invariant Lagrangian torus of F_φ (if ψ_a is a solution of 4.8 so is $\psi_a + c$, $c \in \mathbf{R}$).

(4.11) There exists a dense G_δ set $G \subset \mathbf{A}^n$ such that if $x \in G$, then

$$\sup_{p \in \mathbf{N}} \|p_2(F_\varphi(x))\| = +\infty,$$

where $p_2(\theta, r) = r$.

Using 4.10 we can find a dense G_δ set $G_2 \subset G$, such that when $x \in G_2$, x is recurrent by F_φ (there exists a sequence $p_j \to +\infty$ such that $F_\varphi^{p_j}(x) \to x$).

By 4.11 the diffeomorphism F_φ is not completely integrable (but it is almost by 4.10). The above examples $F_{\varepsilon\varphi}$, $\varepsilon \neq 0$, have very different properties than non completely integrable C^∞ perturbations of a completely integrable monotone diffeomorphism L of $\mathbf{A}$:

• There is no homoclinic phenomenon (when $n = 1$, this follows from the Aubry-Mather theory).

• There are no Birkhoff zones of instability and when $a \in G_1$ there are no compact Aubry Mather sets.

• There are no a priori inequality C^1 or even C^0, for Lagrangian tori invariant by $F_{\varepsilon\varphi}$, $\varepsilon \neq 0$.

It is enough, to see the last points, to choose $(a_i)_i \subset \mathbf{R}^{n-1}$, $\ell_1(a_i) \in DC$, $a_i \to a \in G_1$. Then every sequence of non empty compact sets $K_i \subset M_{a_i}$, invariant by $F_{\varepsilon\varphi}$, is unbounded in $\mathbf{A}$. For, if we suppose that a subsequence of (K_i) is bounded taking a convergent subsequence in the Hausdorff topology, we would conclude that f_a leaves invariant a non empty compact set and therefore has a bounded orbit, hence $a \notin G_1$. This implies that oscillations of the functions $(\psi_{a_i} - \psi_{a_i}(0))$ tends to $+\infty$ when i tends to $+\infty$. If we ask that $\ell_1(a_i) \in \mathbf{Q}$ we conclude that every periodic orbit of $F_{\varepsilon\varphi}$ becomes unbounded when $a_i \to a \in G_1$.

• For every $x \in \mathbf{A}^n$ all the Liapounoff exponents exist at x and are equal to 0.

$\theta_1 = \frac{1}{2}$

The important thing to notice, in the above examples, is that L satisfies :

(4.12) $Q_L(r)(v, v) = {}^t v D\ell(r) v = 0$ for every $r \in \mathbf{R}^n$ with $v = (1, 0, \ldots, 0) \in \mathbf{Q}^n - \{0\}$.

Conjugating L by $T^* A(\theta, r) = (A\theta, {}^t A^{-1} r)$, $A \in GL(n, \mathbf{Z})$, (cf. §3 of the proof of 5.2) it is enough to ask that 4.12 holds with a fixed $v \in \mathbf{Q}^n - \{0\}$ to obtain perturbations of L (we suppose that $L \in C^\omega$) such that the conclusions 4.10 and 4.11 hold.

4.13

Remark. The condition 4.12 is of course highly non generic among completely integrable C^∞ diffeomorphisms L. It violates Nekoroshev's steepness condition [N]. It also shows that the steepness condition enters in Nekoroshev's theorem : we fix $\varepsilon \neq 0$ and consider (using 4.9) $x_0 \in \mathbf{A}^n$ close to $0 \in \mathbf{A}^n$ and $q \in \mathbf{N}$ such that $\|p_2(F^q_{\varepsilon q}(x_0))\| > 10^{10}$. The last property is stable under C^∞ perturbations of L and $F_{\varepsilon\varphi}$.

4.14. Lines of escape

We suppose L is monotone and we can find $a \in \mathbf{R}^{n-1}$ such that for every $r_1 \in \mathbf{R}$

$$ {}^t v \, D\ell(r_1, a)v = 0 \,,\; v = (1, 0, \ldots, 0) \in \mathbf{Q}^n - \{0\} \tag{4.15} $$

or equivalently, if $L(\theta, r) = (\theta + \ell(r), r)$, $\ell(r) = (\ell_1(r), \ldots, \ell_n(r))$, then :

$$ \frac{\partial \ell_1(r_1, a)}{\partial r_1} = 0 \quad \text{for every} \quad r_1 \,. $$

We suppose that φ has the same properties as in 4.7 and we define $F_\varphi = G_\varphi \circ L$ and, for $b \in \mathbf{T}^1$, $F_{\varphi,b} = F_\varphi + (b, 0, \ldots, 0)$. Arguing the same way as we did in 4.9, we can find a dense G_δ set $G_1 \subset \mathbf{T}^1$ such that $b \in G_1$, for every $(\theta, r_1) \in \mathbf{T}^n \times \mathbf{R}$.

$$ \sup_{p \in \mathbf{N}} \|p_2(F^p_{\varphi,b}(\theta, r_1, a))\| = +\infty \tag{4.16} $$

and for almost every $b \in \mathbf{T}^1$ each point $(\theta, r_1, a) \in \mathbf{A}^n$ is on a C^∞ Lagragian torus (that is a graph) invariant by $F_{\varphi,b}$.

The condition 4.16 says that one *can escape to infinity along the line* $s \in \mathbf{R} \to x_0 + sv \in \mathbf{R}^n$, $x_0 = (0, a)$.

Conjugating by translations and by T^*A, $A \in GL(n, \mathbf{Z})$, we can replace 4.15 by the condition $(L = L_{\hat{\ell}})$:

$$ \frac{d^2}{ds^2} \hat{\ell}(x_0 + sv) = 0 \,, \quad \text{for every} \quad s \in \mathbf{R} \tag{4.17} $$

where $x_0 \in \mathbf{R}^n$ and $v \in \mathbf{Q}^n - \{0\}$ fixed, and obtain C^∞ perturbations F of L such that

$$ \sup_{p \in \mathbf{N}} \|p_2(F^p(\theta, x_0 + sv))\| = +\infty \,, \quad \text{for every} \quad s \in \mathbf{R} \quad \text{and} \quad \theta \in \mathbf{T}^n \tag{4.18} $$

4.19

Remarks.

1. What we said in 4.13 also holds replacing the condition 4.12 by 4.17. It occurs in numerical examples for the following reason : if we ask that $\hat{\ell}$ has to belong to the space polynomials of degree $\leq n$, then 4.17 will hold on a dense subset of a non empty open set of such polynomials.

2. For Hamiltonians the example of [H_2] can be reformulated, after conjugacy by translations and by T^*A, $A \in GL(n,\mathbf{Z})$, in the following way : We consider C^∞ Hamiltonians $H(\theta,r) = \hat{\ell}(r) + \varepsilon\varphi(\theta_1)$, with $\varepsilon\varphi'(0) \neq 0$, $\hat{\ell}$ has the form 4.3' and $\ell_1(x_0) = 0$ for some $x_0 \in \mathbf{R}^{n-1}$. Immediate integration with the initial conditions $\theta_1 = 0$ and $r = (r_1, x_0)$ shows that if $\theta_1 = 0$, then $s \to x_0 + sv$, $v = (1,0,\ldots,0)$, is a line of escape for the Hamiltonian flow associated to H (that has the $n-1$ integrals $r_2,\ldots,r_n$ in involution). The requirement $\ell_1(x_0) = 0$ imposes the coordinate invariant condition :

$$\frac{d}{ds}\hat{\ell}(x_0 + sv) = 0\,, \qquad \text{for every} \quad s \in \mathbf{R}\,,$$

where $x_0 \in \mathbf{R}^n$ and $v \in \mathbf{Q}^n - \{0\}$.

3. One can also escape along curves : Let $s \in \mathbf{R} \to u(s) \in \mathbf{R}^n$ be a proper C^∞ embedding and $H(r)$ a C^∞ Hamiltonian such that $\frac{\partial H}{\partial r}(u(s)) = 0$, $s \in \mathbf{R}$. We extend the function $\frac{du}{ds}(s)$ at $u(s)$, $s \in \mathbf{R}$, to a C^∞ function $r \in \mathbf{R}^n \to v(r) \in \mathbf{R}^n$ (i.e. $v(u(s)) = \frac{du}{ds}(s)$, $s \in \mathbf{R}$). We consider the Hamiltonian $H(r) - \frac{\varepsilon}{2\pi}\langle v(r), (\sin 2\pi\theta_1,\ldots,\sin 2\pi\theta_n)\rangle$ and the Hamiltonian flow integrated the initial condition $r = u(0)$ and $\theta = 0$, will escape to infinity along the integral curve $\theta = 0$ and $s \to v(\varepsilon s)$.

In this example it does not seem always clear to the author how to perturb the Hamiltonian such that the escape curve appears as a limit of invariant tori, as was the case of the examples of 4.14.

The reader should consult N.N. Nekoroshev's work [N,1.16C, p.13] and the examples of J. Moser [M_2].

4.20. We suppose that L is monotone, satisfies 4.3, and $f = \mathrm{Id} + \varphi$, is a C^ω orientation preserving diffeomorphism of $\mathbf{T}^1$, where $\varphi_1 \in C^\omega(\mathbf{T}^1,\mathbf{R})$ is such that φ_1 extends analytically to a non constant entire function of $\mathbf{C}$ (e.g. $\varphi_1(\theta_1) = \varepsilon\sin(2\pi\theta_1)$, $0 < \varepsilon < (2\pi)^{-1}$). We define $F = T^*(f \times \mathrm{Id}) \circ L$ where $T^*(f \times I)(\theta,r) = (f(\theta_1),\theta_2,\ldots,\theta_n,(Df(\theta_1))^{-1}r_1,r_2,\ldots,r_n)$; F is a C^ω exact symplectic diffeomorphism of $\mathbf{A}^n$.

PROPOSITION. *There exists an open dense set $U \subset \mathbf{A}^n$ such that, if $x \in U$, then*

$$\lim_{k\to\pm\infty} \|p_2(F^k(x))\| = +\infty\,.$$

Proof. The set $W = \{(r_1,a) \in \mathbf{R}^n \;;\; a \in \mathbf{R}^{n-1},\ \rho(f_{\ell_1(a)}) \in \mathbf{Q}/\mathbf{Z}\}$ contains an open dense set of $\mathbf{A}^n$ where ℓ_1 is defined in 4.3 and ρ is the rotation number of $f_{\ell_1(a)} = \mathrm{Id} + \ell_1(a) + \varphi_1$ (the map $a \to \ell_1(a)$ is a submersion since L is monotone). This follows from [H_4,III.3]. Using again [H_4,II] when $\rho(f_\lambda) = p/q \in \mathbf{Q}/\mathbf{Z}$, $\lambda = \ell(a)$, f_λ has only a finite number periodic points mod 1 (i.e. the fixed points f_λ^q mod 1). The set $U_1 = \{(\theta,r) \in \mathbf{A}^n \;;\; r \in W,\ r_1 \neq 0,\ \theta \notin \{$ periodic points of $f_{\ell_1(r)}\}$ contains an open and dense set $U \subset U_1$. The fact that, when $x \in U_1$, $\lim_{k\to\pm\infty} \|p_2(F^k(x))\| = +\infty$, follows from the following lemma. □

LEMMA. *Let $a < b$ and g be a C^2 diffeomorphism of $[a,b]$, such that $g(a) = a$, $g(b) = b$ and $g(x) \neq x$ if $x \in]a,b[$. Then for every $x \in]a,b[$ we have*

$$\lim_{k \to \pm\infty} (Dg^k)^{-1}(x) = +\infty .$$

Proof. Let $x_0 \in]a,b[$ and $x_1 = g(x_0) \neq x_0$. We denote by I the compact interval bounded by x_0 and x_1. When $k \to \pm\infty$ the length $|g^k(I)|$ of the interval $g^k(I)$ tends to 0. By the means value theorem $Dg^k(\xi_k) = \frac{|g^k(I)|}{|I|}$, $\xi_k \in I$. The lemma follows from the inequality :

$\sup_{z_1, z_2 \in I} |\operatorname{Log} Dg^k(z_1) - \operatorname{Log} Dg^k(z_2)| \leq \operatorname{Var}_{[a,b]}(\operatorname{Log} Dg) =$ total variation of $\operatorname{Log} Dg$ on the interval $[a,b]$. □

4.21. In the above example, the open set U has infinite Haar measure. The global theorem of conjugacy of diffeomorphisms of the circle (see [H$_4$][Y]) implies when $\rho(f_{\ell_1(a)}) \in DC$, $a \in \mathbf{R}^{n-1}$, every (θ, r_1, a) is on a C^ω Lagrangian torus that is a graph. On the nowhere dense set $\mathbf{A}^n - W$, generically the orbits of F are unbounded (this follows from the same arguments as in 4.9 using [H$_4$, IV 4 and 6, XII.2]).

On the set U, F is wandering (F is exact symplectic). Wandering open sets can exist, when $n = 1$, for large perturbations (see [H$_5$, p.50]). When $n \geq 2$, it is unknown to the author if it can happen for C^∞ uniformely small exact symplectic perturbations of a C^∞ completely integrable, monotone positive definite diffeomorphism $L(\theta, r) = (\theta + Br, r)$, $B \in M_n(\mathbf{R})$. But even in this case, Arnold diffusions to infinity can be constructed by the methods of [D$_1$] for some C^∞ exact symplectic perturbations of L.

4.22. One can obtain Hamiltonian flows on $T^*(\mathbf{T}^{2n})$, $n \geq 2$, with similar properties to 4.10 and 4.11 by considering

$$H = \sum_1^n r_j r_{j+n} - \varphi(\theta_1, \ldots, \theta_n)$$

where $\varepsilon > 0$ and $\varphi \in C^\infty(\mathbf{T}^n, \mathbf{R})$ has all its Fourier coefficients non zero.

The same proposition as the one of 4.20 will hold for the Hamiltonian on $T^*(\mathbf{T}^4)$:

$$H = r_1 r_3 + r_2 r_4 + r_1 \varphi_1(\theta_1, \theta_2) + r_2 \varphi_2(\theta_1, \theta_2)$$

when $\varphi = (\varphi_1, \varphi_2) \in C^\infty(\mathbf{T}^2, \mathbf{R}^2)$ belongs to a suitable dense G_δ in the C^∞ topology that I will not describe in this paper, but just hint that it uses M. Peixoto's theorem, see [P, Chap.IV].

4.23. We suppose L satisfies 4.15. We note $\lambda = \ell_1(a)$. Let f be any C^∞ exact symplectic diffeomorphism of $\mathbf{A}$, homotopic to the identity, and $F = (f \times Id_{|\mathbf{A}^{n-1}}) \circ L$. With the same notations as in 4.9, we have $p_a \circ F_{|M_a} = f_\lambda \circ p_a$ where $f_\lambda = f + (\lambda, 0)$. If f tends to the identity in C^∞ compact open topology, then

F tends to L in the C^∞ topology. This does not force anything "tame" for the dynamics of the diffeomorphism f_λ of $\mathbf{A}$:

• The diffeomorphism f_λ can be ergodic for a Haar measure m of $\mathbf{A}$ (see [K], and also [H_7]).

• Given $\delta > 0$, f_λ can leave invariant $\mathbf{A}_\delta$ and has a dense orbit on $\mathbf{A}_\delta$ and even be m ergodic (see [A], and also [F], [H_7]).

• The diffeomorphism can have invariant a pseudo-circle (separating $\mathbf{A}$ into 2 unbounded components) as a minimal set (see [H_1]).

All the above examples are constructed as limits, in the C^∞ topology, of sequences $(g_j)_j$, $g_j = h_j \circ T^* R_{\alpha_j} \circ h_j^{-1}$ where $T^* R_{\alpha_j}(\theta_1, r_1) = (\theta_1 + \alpha_j, r_1)$, $\alpha_j \in \mathbf{T}^1$ and h_j are C^∞ symplectic diffeomorphisms of $\mathbf{A}$.

5. The failure of Birkhoff's theory for perturbations of monotone indefinite completely integrable diffeomorphisms

5.1. The fact that Birkhoff 's theorems do not generalize to perturbations of some completely integrable monotone indefinite diffeomorphisms follows from the examples of 4.7 or 4.20 we already considered, using the same arguments as in [H_3, I.6]. The reader is referred to [H_3, I.4 to 6] for a rapid description of Birkhoff 's theorems (and references) and what one might expect to generalize.

We want to show that Birkhoff 's theorems do not generalize to perturbations of any monotone indefinite completely integrable diffeomorphism , see also 7.12.

We fix L a C^∞ completely integrable diffeomorphism of $\mathbf{A}^n$, $L(\theta, r) = (\theta + \ell(r), r)$ where $\ell(r) = \frac{\partial \hat{\ell}(r)}{\partial r}$ with $\hat{\ell} \in C^\infty(\mathbf{R}^n, \mathbf{R})$. We also fix $r_0 \in \mathbf{R}^n$ and we require that L satisfies : $D\ell(r_0)$ has a non zero isotropic vector vector v. This will be the case when $D\ell(r_0)$ is indefinite (hence is, by definition, non degenerate).

We will also suppose that $n \geq 2$ and we will study in 5.18 the case $n = 1$.

The following theorem shows that Birkhoff 's theorems do not generalize to C^∞ invariant Lagrangian tori for C^∞ perturbations of L. This constrasts with the results of [H_3].

5.2

THEOREM. *We consider L as above and $r_0 \in \mathbf{R}^n$. There exists a sequence $(F_j)_{j \in \mathbf{N}}$ of C^∞ diffeomorphisms of $\mathbf{A}^n$, exact symplectic, having the following properties :*

(5.3) the sequence $(F_j)_j$ converges to L in the C^∞ compact open topology ;

(5.4) the diffeomorphism F_j leaves invariant a C^∞ Lagrangian torus T_j, homotopic to $\{r = 0\}$ and with a Maslov class equal to 0 (for the definition of the Maslov class, the reader is referred to [H_3,II]) ;

(5.5) the restriction of F_j to T_j is C^∞ conjugated to a translation R_{α_j} of $\mathbf{T}^n$ such that $\alpha_j \in CD$;

(5.6) the normal torsion F_j on T_j is non degenerate and indefinite ;

(5.7) the sequence $(T_j)_{j \in \mathbf{N}}$ converges to $\mathbf{T}^n \times \{r_0\}$ in the Hausdorff topology ;

(5.8) for every j, T_j is not a graph over $\mathbf{T}^n$ (i.e. the projection $p_1 : (\theta, r) \to \theta$ is not injective on T_j).

(5.9) Furthermore we can suppose that the tori T_j are such that if T is a C^1 n-torus, C^1 close to T_j, then T is not a graph ; moreover we can ask that the projection p_1 restricted to T_j has, for every j, generic singularities.

Proof of Theorem 5.2

1. Conjugating L by the translation $(\theta, r) \to (\theta, r + r_0)$ we can suppose that $r_0 = 0$.

2. By the hypothesis of 5.1 we can find $v \in \mathbf{R}^n$, $\|v\| = 1$ such that v is an isotropic vector of $D\ell(0)$:

$$ {}^t v D\ell(0) v = 0 \,. $$

We can approximate the vector v by a sequence $(v_j)_j \subset \mathbf{Q}^n$, $\|v_j\| = 1$, $j \in \mathbf{N}$, and we can find a sequence $(U_j)_j \subset SO(n, \mathbf{R})$ satisfying $U_j v_j = v$ and $\|U_j - 1\| \to 0$ as $j \to +\infty$. Let $\ell_j(r) = \frac{\partial \hat{\ell}}{\partial r}(U_j r) = {}^t U_j \ell(U_j r)$ and $L_j(\theta, r) = (\theta + \ell_j(r), r)$. As $L_j \to L$ in the C^∞ topology, when $j \to +\infty$, it is enough to prove the theorem for L_j. We have reduced to prove the theorem supposing that

(5.10) ${}^t v D\ell(0) v = 0$ for a $v \in \mathbf{Q}^n$, $\|v\| \neq 0$.

3. Changing $\hat{\ell}$ to $\sum_1^n \beta_j r_j + \hat{\ell}(r)$ with $\beta_j \to 0$ for $j = 1, \ldots, n$, we can suppose that

(5.11) $\frac{\partial \hat{\ell}}{\partial r}(0) \in \mathbf{Q}^n$.

4. We consider the vector v given by 5.10. Let us show that it is enough to suppose that in 5.10

(5.12) $v = (1, 0, \ldots, 0) = e_1 \in \mathbf{R}^n$.

To see this, we remark that we can find $k \in \mathbf{Q}^*$ such that $kv = (p_1, \ldots, p_n) \in \mathbf{Z}^n - \{0\}$ and the integers $(p_1, \ldots, p_n)$ are relatively prime. Hence we can find $A \in GL(n, \mathbf{Z})$ such that $Akv = (1, 0, \ldots, 0)$. Let

$$ G(\theta, r) = T^* \, {}^t A(\theta, r) = ({}^t A\theta \,,\, A^{-1} r) $$

and we define $\check{L}$ by the equation

$$ \check{L}(\theta, r) = G^{-1} \circ L \circ G(\theta, r) = (\theta + {}^t A^{-1} \ell(A^{-1} r), r) = (\theta + \check{\ell}(r), r) $$

that satisfies ${}^t e_1 D\check{\ell}(0) e_1 = 0$. As G is symplectic and preserves Lagrangian tori that are graphs, after conjugacy by G is enough to prove the theorem supposing 5.12 holds, as well as 5.11.

5. Let $\frac{\partial \hat{\ell}}{\partial r}(0) = (p_1/q_1, p_2/q_2, \ldots, p_n/q_n) \in \mathbf{Q}^n$.

PROPOSITION. *We can suppose that*

(5.13) $p_1/q_1 = 0$.

Proof. We work in the universal covering $\widetilde{\mathbf{A}^n}$. We consider $H(\theta, r) = (q_1\theta, q_1 r)$ that is a conformally symplectic diffeomorphism : $H^*w_n = q_1^2 w_n$ where w_n is the 2 form defined in 2.1. We define $\check{L}$ by the equation

$$L = H^{-1} \circ T^* R_{\alpha_1} \circ \check{L} \circ H$$

with $T^* R_{\alpha_1}(\theta, r) = (\theta + \alpha_1, r)$, $\alpha_1 = (p_1, 0, \ldots, 0)$ and therefore for $\check{L}(\theta, r) = (\theta + \check{\ell}(r), r)$ we have $\frac{1}{q_1}\check{\ell}(q_1 r) + (\frac{p_1}{q_1}, 0, \ldots, 0) = \ell(r)$. The diffeomorphism $\check{L}$ still satisfies 5.12 and we have $\check{\ell}(0) = (0, p_2, \ldots, p_n)$.

If F is a C^∞ exact symplectic diffeomorphism of $\mathbf{A}^n$, then

$$\widetilde{G} = H^{-1} \circ T^* R_{\alpha_1} \circ \widetilde{F} \circ H\,,$$

is defined in the universal covering $\widetilde{\mathbf{A}}^n$ of $\mathbf{A}^n$. It projects onto $\mathbf{A}^n$ to an exact symplectic diffeomorphism G, that commutes with $T^* R_\alpha$ when $\alpha = p/q_1, p \in \mathbf{Z}^n : G$ is a lift to $\mathbf{A}^n$ of F by the finite covering

$$\begin{array}{cccc} G: & \mathbf{A}^n & \longrightarrow & \mathbf{A}^n \\ & \overline{H}\big\downarrow & & \big\downarrow\overline{H} \\ F: & \mathbf{A}^n & \longrightarrow & \mathbf{A}^n \end{array}$$

where $\overline{H}$ is the conformally symplectic finite covering $\overline{H} : (\theta, r) \in \mathbf{A}^n \to (q_1\theta, q_1 r) \in \mathbf{A}^n$.

The completely integrable diffeomorphism $\check{L}$ satisfies 5.11 to 5.13. If one proves the theorem for $\check{L}$ with a sequence $(F_j)_j$ considering the diffeomorphisms obtained by finite covering :

$$H^{-1} \circ T^* R_{\alpha_1} \circ \widetilde{F}_j \circ H = \widetilde{G}_j$$

all the statements of the theorem will hold for L, by finite covering, for the sequence of diffeomorphisms $(G_j)_j$ and furthermore $G_j \to L$ in the C^∞ topology. □

6. We suppose now that L satisfies 5.11 to 5.13 and have to construct the sequence $(F_j)_j$ with the properties 5.3 to 5.9.

Let $\ell_1(r) = \frac{\partial \ell}{\partial r_1}(r)$; we can always suppose that $\hat{\ell}(0) = 0$. It follows from 5.13 that $\ell_1(0) = 0$ and from 5.11 $\frac{\partial^2 \ell}{\partial r_1^2}(0) = 0$, hence $\hat{\ell}(r_1, 0) = O(r_1^3)$. Perturbing $\hat{\ell}$ to $\hat{\ell}(r) + \varepsilon r_1^3$ we can suppose that

$$\text{(5.14)} \qquad \hat{\ell}(r_1, 0) = a r_1^3 + O(r_1^4) \quad \text{when} \quad r_1 \to 0, \quad \text{with} \quad a \neq 0$$

We can write, when r_1 is small,

$$\text{(5.15)} \qquad \hat{\ell}(r_1, 0) = a(g(r_1))^3 \circ g(r_1), \text{ when } r_1 \to 0, \text{ where } g(r_1) = r_1 + O(r_1^2)$$

is a C^∞ local diffeomorphism near $r_1 = 0$.

We consider the Hamiltonian

(5.16) $H_\varepsilon(\theta,r) = \hat{\ell}(r) - \varepsilon(\cos 2\pi(\theta_1))\eta(r)$, $\varepsilon > 0$ and $X_{H_\varepsilon} = (\frac{\partial H}{\partial r}, -\frac{\partial H}{\partial \theta})$ the Hamiltonian vector field associated to H_ε, where η is a C^∞ function with the properties

$$\begin{cases} \eta(r) = 1\,, & \text{if } \|r\| < 1, \\ \text{and} & \\ \eta(r) = 0\,, & \text{if } \|r\| > 100. \end{cases}$$

We denote by F_ε the time one on the complete flow of the vector field X_{H_ε}.

Even if we replace the function η by $\eta \equiv 1$, since $|\frac{dr_1}{ds}| \le 2\pi\varepsilon$, the flow of X_{H_ε} is complete. We use the function η to obtain 5.3 in the C^∞ fine Whitney topology.

When $\varepsilon \to 0$, the diffeomorphisms $F_\varepsilon \to L$ in the C^∞ topology.

The Hamiltonian H_ε is "completely integrable" in angle action coordinates since it has the integrals in involution $r_2, \ldots, r_n$ and $H_\varepsilon = \delta$, $\delta \in \mathbf{R}$; but these integrals have singularities. When $0 < |\varepsilon| < \varepsilon_0$, F_ε leaves invariant the Lagrangian torus T_ε graph of

$$\theta \in \mathbf{T}^n \to \left(g^{-1}\left(\left(\frac{\varepsilon}{a}\cos 2\pi\theta_1\right)^{1/3}\right), 0, \ldots, 0\right) \in \mathbf{R}^n .$$

The number ε_0 depends on a and g.

When ε is small enough, $\varepsilon \neq 0$, T_ε is a C^∞ sub-manifold of $\mathbf{A}^n$, but T_ε has a vertical tangent vector at $\theta_1 = \frac{1}{4}$ or $\frac{3}{4}$: the Hamiltonian on $\mathbf{A}$:

$$\widehat{H}_\varepsilon(\theta_1, r_1) = \hat{\ell}(r_1, 0, \ldots, 0) - \varepsilon\eta(r_1, 0, \ldots, 0)\cos 2\pi\theta_1$$

has no singular points on T_ε.

7. Perturbing H_ε by $kH_\varepsilon + \sum_2^n \beta_i r_i$ with $k \to 1$ and $\beta_i \to 0$ for $i = 2, \ldots, n$, we can find a sequence $(\varepsilon_j)_{j\in\mathbf{N}} \subset \mathbf{R}_+^*$ such that $F_{\varepsilon_j|T_{\varepsilon_j}}$ is C^∞ conjugated to $\alpha_j \in CD$ where $\varepsilon_j \to 0$, $k = k_j \to 1$ and $\beta_i = \beta_{i,j} \to 0$, $2 \le i \le n$ as $j \to +\infty$ (we take sequences since the set $DC \subset \mathbf{T}^n$ is totally disconnected).

The fact that $F_{\varepsilon_j|T_{\varepsilon_j}}$ is C^∞ conjugated to a translation follows either from the next step 8 of the proof of 5.2 or the following elementary lemma that requires no small divisors and that we leave the proof to the reader.

LEMMA. *Let X be a C^∞ vector fields on $\mathbf{T}^n$ that depends only on the first coordinate θ_1 and such that the first component of X_1 of X, is non zero at every point θ_1. Then X is C^∞ conjugated to a constant vector field on $\mathbf{T}^n$.*

When $\varepsilon_j \to 0$, 5.7 is satisfied.

As $\varepsilon_j \neq 0$, since F_{ε_j} leaves invariant a torus T_{ε_j} with a vertical tangent vector, one can find a sequence of C^∞ exact symplectic diffeomorphisms $(K_{\varepsilon_j})_j$, $K_{\varepsilon_j} \to \mathrm{Id}$ in the C^∞ topology and such that for $F_j = K_{\varepsilon_j}^{-1} \circ F_{\varepsilon_j} \circ K_{\varepsilon_j}$, that leaves invariant $T_j = K_{\varepsilon_j}^{-1}(T_{\varepsilon_j})$, 5.9 holds for the tori T_j.

All the conclusions of 5.2 with the possible exception of 5.6 hold (by 3.9 is enough to perturb F_{ε_j} leaving invariant T_{ε_j} and then consider $K_{\varepsilon_j}^{-1} \circ F_{\varepsilon_j} \circ K_{\varepsilon_j}$).

8. We have to show how to obtain examples with the property 5.6. We can always, using 3.9, ensure that the normal torsion of F_{ε_j} on T_{ε_j} is non degenerate.

In fact we can remark that the Hamiltonian H_ε has angle action variables near T_ε : we make a C^ω symplectic change of variables of the form

$$G\ :(\theta_1,r_1,\theta_2,r_2\ldots\theta_n,r_n)\to(G_1(\theta_1,r_1)\,,\,\theta_2,r_2,\ldots,\theta_n,r_n),$$

such that $G^{-1}(T_\varepsilon)$ is the graph of a C^∞ function. The conjugated Hamiltonian $H_\varepsilon \circ G$ still has the functions $r_2,\ldots,r_n$ as integrals in involution and there are no singularities near $G^{-1}(T_\varepsilon)$. In the angle action variables, we can perform the perturbations of 3.9, keeping the fact that F_ε is completely integrable, in angle action variables, near T_ε.

9. We still have to show how to prove the second part 5.6 : the torsion can be made indefinite.

We have to change the construction of the tori T_ε. We suppose that $\varepsilon > 0$ is small and $-2\varepsilon < \delta < 2\varepsilon$. The diffeomorphism F_ε leaves invariant the tori $C_{\varepsilon,\delta}\times\{0\}\subset \mathbf{A}\times\mathbf{A}^{n-1}$ where $C_{\varepsilon,\delta}=\Gamma_{\psi_{\varepsilon,\delta}}\subset\mathbf{A}$ and

$$\psi_{\varepsilon,\delta}(\theta_1)=g^{-1}\left(\left(\frac{\varepsilon}{a}(\cos 2\pi\theta_1)+\frac{\delta}{a}\right)^{1/3}\right).$$

The one dimensional tori $C_{\varepsilon,\delta}$ (also called curves) are invariant by the Hamiltonian flow associated to the Hamiltonian $H_\varepsilon(\theta,r_1,0,\ldots,0)=\widehat{H}_\varepsilon(\theta_1,r_1)$ on $\mathbf{A}$. We denote by G_ε the time one of the exact symplectic flow associated to the Hamiltonian $\widehat{H}_\varepsilon$ on $\mathbf{A}$. When $-\varepsilon<\delta<\varepsilon$, $C_{\varepsilon,\delta}$ is a C^∞ sub-manifold ($\widehat{H}_\varepsilon$ has no singular points on $C_{\varepsilon,\delta}$) and has vertical tangents at 2 different points. If $\delta=-\varepsilon,\varepsilon$ $C_{\varepsilon,\pm\varepsilon}$ are topological circles (they each have a cusp singularity) and G_ε has a fixed point on each circle $C_{\varepsilon,\delta}$, $\delta=\pm\varepsilon$.

The function $\delta\in[-\varepsilon,\varepsilon]\to a\rho(G_{\varepsilon|C_{\varepsilon,\delta}}))=\hat\rho(\delta)$ is continuous, C^∞ on $]-\varepsilon,\varepsilon[$, where the number a is defined in 5.14 and ρ denotes the rotation number of the homeomorphism G_ε on the circle $C_{\varepsilon,\delta}$. We have $\hat\rho(-\varepsilon)=\hat\rho(\varepsilon)=0$ and $\hat\rho(\delta)>0$ when $\delta\in]-\varepsilon,\varepsilon[$ (it is enough to draw, with arrows, the level lines of $\hat H_\varepsilon$). We therefore can find at least one $\delta(\varepsilon)\in]-\varepsilon,\varepsilon[$ such that $\hat\rho(\delta(\varepsilon))$ is a maximum on $[-\varepsilon,\varepsilon]$, hence $\frac{d}{d\delta}\ \hat\rho(\delta(\varepsilon))=0$. The normal torsion of G_ε on $C_{\varepsilon,\delta}$ is defined when $\delta\in]-\varepsilon,\varepsilon[$ (G_ε is on a flow and $G_{\varepsilon|C_{\varepsilon,\delta}}$ is C^∞ conjugated to a rotation). The normal torsion, calculated by using the step 8 of the proof of 5.2 and 3.5, is a C^∞ function of $\delta\in]-\varepsilon,\varepsilon[$ (one uses locally angle action variables). We have :

(5.17) The normal torsion of G_ε on $C_{\varepsilon,\delta(\varepsilon)}$ is equal to 0.

This follows from the fact that 2 different families of invariant circles with the same rotation numbers collide at $\delta=\delta(\varepsilon)$ (if $\hat\rho^{-1}(\hat\rho(\delta(\varepsilon)))=J$ has non empty interior, we can always suppose that $\delta(\varepsilon)\in\mathrm{Int}(J)$).

This fact stays true, for the same $\delta=\delta(\varepsilon)$, if we replace H_ε by kH_ε, $k\in\mathbf{R}^*$.

We now replace the invariant tori T_ε we considered in the steps 6 to 8 of the proof of 5.2 by the tori $T_{\varepsilon,\delta(\varepsilon)}$ ($T_\varepsilon=T_{\varepsilon,0}$). Everything we stated and did for T_ε still applies to the tori $T_{\varepsilon,\delta(\varepsilon)}$ (we use $|\delta(\varepsilon)| < \varepsilon$). We want to calculate the normal torsion of F_ε on $T_{\varepsilon,\delta(\varepsilon)}$. To do this, we use the step 8 and 3.5. The way we proved, 5.17, still applies and this forces the normal torsion B_ε of G_ε on $T_{\varepsilon,\delta(\varepsilon)}$ to be such that $b_{11,\varepsilon} = 0$ where $B_\varepsilon = (b_{ij,\varepsilon})$, i.e. $B_\varepsilon = \begin{pmatrix} 0 & * & * \\ * & * & * \\ * & * & * \end{pmatrix}$. The symmetric matrix B_ε is either degenerate or indefinite. If B_ε is indefinite, 5.6 holds, otherwise we can perturb B_ε to $B_\varepsilon + \Delta B_\varepsilon$ such that ΔB_ε is symmetric, $\|\Delta B_\varepsilon\| \to 0$, $\det(B_\varepsilon + \Delta B_\varepsilon) \neq 0$ and $B_\varepsilon + \Delta B_\varepsilon$ is indefinite. To ensure 5.6 we now proceed as in the step 8 above. The theorem 5.2 follows from the steps 1 to 9. □

5.18. We consider the case $n = 1$. We ask that $L : \mathbf{A} \to \mathbf{A}$ is C^∞, of the form $L(\theta, r) = (\theta + \ell(r), r)$ and L has 0 torsion on $\mathbf{T}^1 \times \{r_0\}$, i.e. $\frac{d\ell}{dr}(r_0) = 0$.

PROPOSITION. *Given a diffeomorphism L as above then we can construct a sequence $(F_j)_{j\in\mathbf{N}}$ of exact symplectic diffeomorphisms of $\mathbf{A}$ such that 5.3 to 5.9 hold replacing 5.6 by :*

(5.6′) *The normal torsion of F_j on T_j is equal to 0.*

The proof is exactly the same as for 5.2 (the steps 2 and 4 are not needed and 5.17 is used).

Furthermore we can use the step 8 of the proof of 5.2.

5.19. The diffeomorphism F_j is C^∞ conjugated on a neighbourhood of T_j, by a symplectic diffeomorphism , to a C^∞ completely integrable diffeomorphism .

5.20

Remarks.

1. We can replace, in 5.2, C^∞ by C^ω. We have to impose in this case that $L \in C^\omega$ and F_j are C^ω exact symplectic diffeomorphisms of $\mathbf{A}^n$ (one should take in 5.16, $\eta \equiv 1$). The only property that might not hold (or at least the proof given does not work) is the fact that in 5.6 the normal torsion of F_j on T_j is non degenerate.

2. In any C^∞ neighbourhood of L one can repeat infinitely often the constructions 5.2 and 5.18, using for example what we said in the step 8 of the proof of 5.2 or 5.19.

5.21. We perturb the Hamiltonian H_ε to $H_{\varepsilon,b}(\theta, r) = H_\varepsilon(\theta, r) + b\hat{g}(r_1)$ where $\hat{g}$ is a C^∞ function, equal to the diffeomorphism $g(r_1)$ when $|r_1|$ is small enough, where g is defined in 5.15 and $b \in \mathbf{R}$. We suppose that $ba < 0$ and $\frac{b}{a}$ tends to 0. The Hamiltonian $\widehat{H}_{\varepsilon,b}(\theta_1, r_1) = \widehat{H}_\varepsilon(\theta_1, r_1) + b\hat{g}(r_1)$ has 4 critical points near $\{r_1 = 0\}$. We obtain the level lines of $\widehat{H}_{\varepsilon,b}$ in $\mathbf{A}$, near $\{r_1 = 0\}$, where ε is small and $b \to 0$ as represented in the following picture :

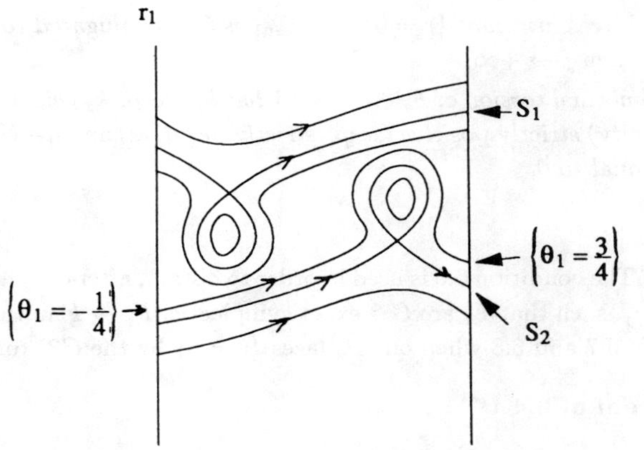

Figure: **Level lines of $\widehat{H}_{\varepsilon,b}$, $a > 0$.**

We have 2 singular curves (each is the image of a continuous immersion of $\mathbf{T}^1$), S_1 and S_2. Both S_j are homeomorphic to the wedge $\mathbf{T}^1 \vee \mathbf{T}^1$ and are limits of C^∞ invariant circles of the Hamiltonian flow associated to $\widehat{H}_{\varepsilon,b}$. For the flow associated to the Hamiltonian $H_{\varepsilon,b}$ we obtain that C^∞ invariant Lagrangian tori T_j can converge in the Hausdorff topology to $S_1 \times \mathbf{T}^{n-1}$ that is not a manifold. On T_j we can suppose that 5.5 holds and the normal torsion is non degenerate (see 3.9). We can then perform the steps to 1 to 9 of the proof of 5.2.

6. Changing the normal torsions by C^∞ perturbations

6.1. We suppose that F is a C^∞ symplectic diffeomorphism of $\mathbf{A}^n$ that leaves invariant a C^∞ Lagrangian torus T and $F_{|T}$ is C^∞ conjugated to R_α, $\alpha \in DC$. Then using what we said in 3.4 and the Birkhoff normal form 3.8 (with $k = 1$) we can suppose after conjugacy of F, that for a $\delta > 0$

(6.2) $F : \mathbf{A}^n_\delta \to \mathbf{A}^n$ is a C^∞ exact symplectic embedding that leaves invariant $\{r = 0\} \equiv \mathbf{T}^n_0$

and

(6.3) $F(\theta, r) = (\theta + \alpha + Br + O(r^2), r + O(r^2))$

when $r \to 0$. The matrix B is symmetric and its type does not depend on the choice of coordinates (see 3.4).

6.4

THEOREM. *We suppose that F satisfies 6.2, 6.3 and the matrix B is indefinite. We fix non negative integers k_1 and k_2 such that $1 \le k_1 + k_2 \le n$, then there exists a sequence $(F_j)_{j\in\mathbf{N}}$ of C^∞ exact symplectic embeddings of $\mathbf{A}^n_\delta$ into $\mathbf{A}^n$ that satisfy :*

(6.5) $F_j(\theta, r) = F(\theta, r)$, if $\|r\| \ge \delta/2$;

(6.6) $F_j \to F$ in the C^∞ topology as $j \to +\infty$;

(6.7) F_j leaves invariant $\{r=0\}$, $F_{j|\{r=0\}}$ is C^∞ conjugated to R_{α_j}, $\alpha_j \in DC$ and $\alpha_j \to \alpha$ when $j \to +\infty$,

(6.8) the normal torsion of F_j on $\{r=0\}$ has k_1 (resp. k_2) eigenvalues (counted with multiplicity) strictly positive (resp. strictly negative) and $n-(k_1+k_2) \leq n-1$ eigenvalues equal to 0.

6.9

Remark. The condition 6.6 is used in order to obtain, after conjugacy (cf. 6.1), a sequence $(F_j)_j$, such that F_j are C^∞ exact symplectic, $F_j \to F$ in the C^∞ topology and F_j satisfy 6.7 and 6.8 when one replaces $\{r=0\}$ by the C^∞ torus T.

6.10. Proof of 6.4

6.11. Replacing F by $F \circ L_j$, where L_j, $j \in \mathbf{N}$ are C^∞ completely integrable diffeomorphisms , $L_j(\theta, r) = (\theta + \ell_j(r), r)$ such that $\ell_j(r) = 0$ if $\|r\| \geq \delta/2$ and $L_j \to \mathrm{Id}_{\mathbf{A}^n}$ in the C^∞ topology as $j \to +\infty$, we can suppose that F satisfies 6.3 with :

(6.12) $\alpha \in \mathbf{Q}$, $\alpha = p/q$, $p \in \mathbf{Z}^n$, $q \in \mathbf{N}^*$.

and

(6.13) $B \in M_n(\mathbf{Q})$.

We will use the following well known lemma.

6.14

LEMMA. *The group $SO(n, \mathbf{Q})$ is dense in $SO(n, \mathbf{R})$.*

Proof. If $A \in M_n(\mathbf{R})$ is antisymmetric, the Cayley's transform of A satisfies

$$K_A = (1+A)(1-A)^{-1} \in SO(n, \mathbf{R})$$

and the map $A \to K_A$ is a local diffeomorphism of a neighbourhood of 0 onto a neighbourhood of $K_0 = 1$ (1 denotes the unit matrix).

Let $\mathcal{A}_{\mathbf{Q}} = \{A \in M_n(\mathbf{Q})\,;\, {}^tA = -A\}$. If $A \in \mathcal{A}_{\mathbf{Q}}$, then $K_A \in SO(n, \mathbf{Q})$. As $\mathcal{A}_{\mathbf{Q}}$ is dense in $\mathcal{A}_{\mathbf{R}}$, it follows that $SO(n, \mathbf{Q})$ is dense in a neighbourhood of 1 in the connected topological group $SO(n, \mathbf{R})$ and this implies the lemma. □

It follows from the above lemma and the same argument as in 6.11 that we are reduced to prove 6.4 replacing 6.13 by

(6.15) $B = U \Delta_1 {}^tU$, $\Delta_1 \in M_n(\mathbf{Q})$ is a diagonal matrix and $u \in SO(n, \mathbf{Q})$.

Let $q_1 \in \mathbf{N}^*$ such that ${}^tA = q_1 u \in M_n(\mathbf{Z})$.

We have $\det(q_1 u) = q_1^n$ and

(6.16) $$AB\,{}^tA = \Delta = q_1^2 \Delta_1 = \begin{pmatrix} a_1 & & 0 \\ & \ddots & \\ 0 & & a_n \end{pmatrix}.$$

6.17. We look for $f_1 = h \circ R_{\beta_1} \circ h^{-1}$, $\beta_1 \in DC$ $h \in D^\infty(\mathbf{T}^n)$, where $D^\infty(\mathbf{T}^n)$ denotes the group of diffeomorphisms obtained by lifting to $\mathbf{R}^n$, C^∞ diffeomorphisms of $\mathbf{T}^n$, homotopic to the identity :

$$D^\infty(\mathbf{T}^n) = \{f \in \text{Diff}^\infty(\mathbf{R}^n),\ f - \text{Id} \quad \text{is} \quad \mathbf{Z}^n - \text{ periodic }\}.$$

Then we consider $f = A^{-1} \circ f_1 \circ A \in D^\infty(\mathbf{T}^n)$ thus

$$f = A^{-1} \circ h \circ A \circ R_\beta \circ A^{-1} \circ h^{-1} \circ A \quad \text{with} \quad A\beta = \beta_1.$$

Let $G = F \circ T^* R_{-p/q} \circ T^* f$ where $T^* f(\theta, r) = (f(\theta), {}^t(Df(\theta))^{-1} r)$ and $\alpha = p/q$ is defined in 6.12. As the diffeomorphism $T^* f$ is symplectic, G is a C^∞ symplectic embedding of neighbourhood of $\{r = 0\}$ into $\mathbf{A}^n$ (we explain later on how to obtain 6.2), that leaves invariant $\{r = 0\}$. We have

$$G(\theta, r) = (f(\theta) + B^t(Df)^{-1}(\theta) r + O(r^2),\ {}^t(Df(\theta))^{-1} r + O(r^2))$$

as $r \to 0$. We calculate the normal torsion B_G of G on $\{r = 0\}$ as in 3.3 :
Let $H = T^*(A^{-1} \circ h \circ A)$ we consider $H^{-1} \circ G \circ H$ and obtain

$B_G = A^{-1} C\, {}^t A^{-1}$,

$C = \int_{\mathbf{T}^n} ((Dh)^{-1} \circ A \circ R_\beta)\ AB\, {}^t A\, {}^t (Dh \circ A \circ R_\beta)^{-1} d\theta$,

hence using 6.15

(6.18) $\quad C = \int_{\mathbf{T}^n} (Dh)^{-1}(\theta) \Delta\, {}^t (Dh)^{-1}(\theta) d\theta$,

since the continuous affine map $A \circ R_\beta : \mathbf{T}^n \to \mathbf{T}^n$ preserves the Haar measure.

If $B_1 \in M_n(\mathbf{R})$ is a symmetric matrix, we denote by $k_+(B_1)$ (resp. $k_-(B_1)$) the number of eigenvalues, counted with multiplicity, that are strictly positive (resp. strictly negative).

We note $A\frac{p}{q} = p_2/q_2$, $p_2 \in \mathbf{Z}^n$ and $q_2 \in \mathbf{N}^*$ where $\frac{p}{q} = \alpha$ is given in 6.12.

We will suppose that the number k_2 in 6.8 is > 0 (when $k_1 > 0$ the proof is the same by changing + signs into - signs, or replacing F by F^{-1}).

6.19

LEMMA. *We can find $h \in D^\omega(\mathbf{T}^n)$ such that*

(6.20) $\quad h \circ R_{p_2/q_2} = R_{p_2/q_2} \circ h$

and

(6.21) $\quad k_+(B_G) = n - 1, \quad k_-(B_G) = 1.$

Proof. As the matrix B is indefinite in the matrix Δ of 6.15 we can find a_j and a_k of opposite sign. Conjugating h by permutation matrix (i.e. in $O(n, \mathbf{Z})$), we can suppose $a_1 > 0$ and $a_n < 0$.

We take $\quad h(\theta_1, \ldots, \theta_n) = (\theta_1, \theta_2 - \varphi_2(\theta_1), \ldots, \theta_n - \varphi_n(\theta_1))$.

We suppose that

$$\int_{\mathbf{T}^1} \varphi_j'(\theta_1) \varphi_k'(\theta_1) d\theta_1 = 0\,,\ 2 \le k < j,$$

and for $2 \leq j \leq n$, φ_j is $\frac{1}{q_2}$-periodic (e.g. $\varphi_j(\theta_1) = c_j \sin(2\pi j\, q_2\theta_1)$, $c_j \in \mathbf{R}$).

We calculate the matrix C of 6.18 and obtain

$$C = \ \mathrm{diag}\,\Big(a_1, a_2 + a_1 \int_{\mathbf{T}^1} (\varphi_2'(\theta_1))^2 d\theta_1, \ldots, a_n + a_1 \int_{\mathbf{T}^1} (\varphi_n'(\theta_1))^2 d\theta_1\Big).$$

Given numbers $b_j \geq a_j$, $j \geq 2$, we can find h as above such that

$$a_j + a_1 \int_{\mathbf{T}^1} (\varphi_j'(\theta_1))^2 d\theta_1 = b_j, \qquad \text{for every} \quad j \geq 2.$$

The number b_j can be < 0, $= 0$, or > 0 when $a_j < 0$ (since $a_1 > 0$).

As $B_G = A^{-1} C\, {}^t\!A^{-1}$ (using the invariance of the signature of a quadratic form or the fact that $\frac{1}{q_1} A \in SO(n, \mathbf{Q})$) we can construct $h \in D^\omega(\mathbf{T}^n)$ that satisfies 6.20 and 6.21. □

6.22

LEMMA. *We suppose that $B = B_F$ satisfies*

$$k_+(B_F) = n - 1\,,\ k_-(B_F) = 1\,. \tag{6.23}$$

Given the numbers k_1 and k_2 of 6.8 ($k_2 > 0$) we can find $h \in D^\infty(\mathbf{T}^n)$ such that

$$h \circ R_{p_2/q_2} = R_{p_2/q_2} \circ h \tag{6.24}$$

and

$$k_+(B_G) = k_1\ ,\ k_-(B_G) = k_2\,. \tag{6.25}$$

Proof. Using 6.23 and conjugating by h a permutation matrix we can suppose $a_1 < 0$ and $a_j > 0$, $j \geq 2$. The rest of the proof is the same as for 6.19. □

6.26. The diffeomorphisms h constructed in 6.19 and 6.22 will not usually be C^0 close to the identity. Let us show how to overcome this difficult : it follows from 6.20 and 6.24 that

$$A^{-1} \circ h \circ A \circ R_{p/q} \circ A^{-1} \circ h^{-1} \circ A = R_{p/q}\,.$$

If the sequence $(\alpha_j)_{j \in \mathbf{N}} \subset \mathbf{T}^n$ tends to p/q, as $j \to +\infty$, then

$f_j = A^{-1} \circ h \circ A \circ R_{\alpha_j} \circ A^{-1} \circ h^{-1} \circ A$ tends to $R_{p/q}$ in the C^∞ topology. We can always suppose $(\alpha_j)_j \subset DC$, $\alpha_j \to p/q$, and then we define

$$F_j = F \circ T^* R_{-p/q} \circ T^* f_j$$

in neighbourhood of $\{r = 0\}$ and using 3.9 we can extend $T^* R_{-p/q} \circ T^* f_j$ to be identity on $\{(\theta, r); \|r\| \geq \delta/2\}$. Hence one obtains a sequence F_j that satisfies 6.5 to 6.7.

6.27. The end of the proof

One first constructs a sequence $(F_j)_{j\in\mathbf{N}}$ such that 6.5 to 6.7 hold as well as the conclusions of 6.8 with $k_2 = 1$ and $k_1 = n-1$. This is possible using everything we done , and in particular 6.19 and 6.26. Then for each F_i we have 6.7, as $\alpha_i \in DC$ we can make the changes of coordinates we explained in 6.1 using the Birkhoff normal (the number $\delta = \delta_i$ of 6.2, depends on i and we really need 6.5 in order to apply 6.9). We start over the proof 6.4 with $\delta = \delta_i'$ and $F = F_i$. This time we can use 6.22 (6.23 is satisfied for each F_i) and obtain sequence $(F_{i,j})_{j\in\mathbf{N}}$ that satisfies 6.5 to 6.8 and the theorem follows using 6.9. □

6.28

Remarks.

1. The fact that is behind, the proof is the following :

If $F(\theta, r) = (\theta + \frac{p}{q} + Br + O(r^2), r + O(r^2)))$, when $r \to 0$ and $\frac{p}{q} \in \mathbf{Q}^n$, then $R_{p/q}$ has a large centralizer in $D^\infty(\mathbf{T}^n)$. When B is indefinite and one conjugates F by T^*h, h commutes with $R_{p/q}$, then normal torsion of $T^*h^{-1} \circ F \circ T^*h$ on $\{r = 0\}$ is equal to

$$\int_{\mathbf{T}^n} (Dh)^{-1} B\, {}^t(Dh)^{-1} d\theta$$

and its type is not well determined. If B is definite, the above integral is definite.

2. In 6.4 we can replace C^∞ by C^ω if one does not require 6.5 to hold and in 6.8 we only ask what can be obtained by the method of proof of 6.19 once (e.g. $k_1 = n$, $k_2 = n$ or $k_1 + k_2 \geq n-1$).

The condition 6.5 can be replaced by : F_j is an C^ω exact symplectic embedding defined on $\mathbf{A}^n_{\delta'}$, $\frac{\delta}{2} < \delta' < \delta$. In 6.11, one should not ask that $\ell(r) = 0$, if $\|r\| \geq \delta/2$ since we want $L \in C^\omega$.

6.29. The following consequence shows that for an open set of C^∞ perturbations of monotone indefinite completely integrable diffeomorphisms the uniqueness of invariant Lagrangian diophantine tori, with a fixed rotation vector in the universal cover, strongly fails. This contrasts with the monotone positive (definite) case (see [H$_3$, I.10]).

Consequence : *Given $r_0 \in \mathbf{R}^n$ and L a C^∞ completely integrable diffeomorphism such that its torsion on $\{r_0 = 0\}$ is indefinite. Then there exists a sequence $(F_j)_{j\in\mathbf{N}}$ of C^∞ exact symplectic diffeomorphisms satisfying :*

- *$(F_j)_{j\in\mathbf{N}}$ tends to L in the C^∞ topology, as $j \to +\infty$;*
- *for every j, F_j leaves invariant two C^∞ Lagragian tori T_j and T_j', graphs of C^∞ functions ψ_j and ψ_j' in $C^\infty(\mathbf{T}^n, \mathbf{R}^n)$;*
- *when $j \to +\infty$, ψ_j and ψ_j' tend, in the C^∞ topology, to the constant function r_0 ;*
- *both $F_{|T_j}$ and $F_{|T_j'}$ are C^∞ conjugated to R_{α_j}, $\alpha_j \in DC$ (with the same α_j) ;*
- *the normal torsions of F on T_j and T_j' are non degenerate ;*

- *in the universal cover* $\widetilde{F}_{|\widetilde{T}_j}$ *and* $\widetilde{F}_{|\widetilde{T}'_j}$ *have the same rotation vector (defining rotation vectors as in [H₃, I.10.2]) ;*

and finally

- *for every* j, $T_j \cap T'_j = \emptyset$.

Idea of the proof. One reduces to the case $r_0 = 0$, of step 1 of the proof of 5.2. Using 6.4 we can perturb L to F in the C^∞ topology such that 6.7 and 6.8 hold and in 6.8 we ask that $k_1 + k_2 = n - 1$ (i.e. the normal torsion has a one 0 eigenvalue).

Given any $k' \in \mathbf{N}^*$, we perform the Birkhoff normal form 3.8 with $k \gg k'$. This implies that : in any neighbourhood V of F in $C^{k'}(\mathbf{A}^n, \mathbf{A}^n)$, with $C^{k'}$ topology, we can find a C^∞ exact symplectic diffeomorphism G, equal to L_k in a small neighbourhood N_V of $\{r = 0\}$ where L_k is given by 3.8. The normal torsion of L_k on $\{r = r_0\}$ is equal to the one F hence has one eigenvalue equal to 0.

Using the same construction as in 6.11 (and [G₂], in order to control the generic singularities) after perturbation, we can suppose that the completely integrable C^∞ diffeomorphism L_k has degenerate normal torsion on $\{r = r_0\}$ and is generic with this property. A *doubling bifurcation of the torus* $\{r = r_0\}$ will occur : we can find 2 sequences $(\tau_j)_j$ and $(\tau'_j)_j$ of different C^∞ Lagrangian tori, graphs of constant functions, satisfying $\tau_j \cap \tau'_j = \emptyset$, $\tau_j \to \{r = r_0\}$ and $\tau'_j \to \{r = r_0\}$ in the C^∞ topology, as $i \to +\infty$, and such that $\widetilde{L}_{k|\tilde{\tau}_j}$ and $\widetilde{L}_{k|\tilde{\tau}'_j}$ have the same rotation vector in the universal cover $\widetilde{\mathbf{A}}^n$ of $\mathbf{A}^n$. We perform the same construction that we did in 6.11 (or in 3.9) to each to tori τ_j and τ'_j, keeping the same rotation vector in the universal cover, we can suppose that the tori τ_j and τ'_j have the same properties we want for the tori T_j and T'_j. Conjugating back, the consequence follows from the fact that the C^∞ topology is the projective limit topology of the $C^{k'}$ topologies.

7. Some generic properties

7.1. We denote by $\mathcal{S}_n$ the group of C^∞, exact symplectic, diffeomorphisms of $\mathbf{A}^n$, homotopic to the identity. With the C^∞ compact open topology, $\mathcal{S}_n$ is a Polish topological group (cf. [H₆, 5.1]). Every open set of $\mathcal{S}_n$ is a Polish topological space, hence is a Baire space.

7.2. We fix in the following two real numbers $\delta_1 > 0$ and $\delta_2 > 0$.

We denote by $IT^\infty(F)$ the set of tori $\tau \subset \mathbf{A}^n$ that satisfy :

- τ is a C^∞ Lagrangian torus homotopic to $\{r = 0\}$;
- τ is invariant by F ;
- $F_{|\tau}$ is C^∞ conjugated to a translation of $\mathbf{T}^n$ that satisfies diophantine condition ;
- the normal torsion of F on τ is non degenerate ;
- $\tau \subset \text{Int}\,(\mathbf{A}^n_{\delta_1}) = \{(\theta, r); \|r\| < \delta_1\}$.

The set $IT^\infty(F)$ is all the invariant tori of F one can hope to ever obtain using 3.4 and 3.6. Requiring that the normal torsion is non degenerate is not very severe if one refers to 3.9 and the last condition is simply made in order to avoid to consider tori that become unbounded (as in 4.10) for otherwise we would have to study the behaviour of F at infinity.

We consider $IT^\infty(F)$ as a subset of the space K of compact subsets of $\mathbf{A}^n_{\delta_1}$. The space K with the Hausdorff topology is a compact metric space, see [M₁,1.10].

We consider the subsets of $IT^\infty(F)$:

$IT^\infty_d(F) = \{\tau \in IT^\infty \text{ ; the normal torsion of } F \text{ on } \tau \text{ is definite}\}$

and

$IT^\infty_i(F) = \{\tau \in IT^\infty \text{ ; the normal torsion of } F \text{ on } \tau \text{ is indefinite}\}$.

We denote by $IT_i(F)$ and $IT_d(F)$ the closure of these sets in K with induced Hausdorff topology. We define $V_i = \{F \in \mathcal{S}_n,\ IT^\infty_i(F) \neq \emptyset\}$ and, in the same way, V_d.

7.3. It follows from 3.4 and 3.6 that the set V_i and V_d are open (and not empty) in $\mathcal{S}_n$. It also implies that the following set valued functions $V_* \to IT^\infty_*(F) \subset K$ and $V_* \to IT_*(F) \subset K$, where $* = i$ or d, are lower semi-continuous, the reader is referred to [H₆,12.3].

7.4. In [H₆,5] we studied the set $Y^\infty(F) = \{\tau \in IT^\infty(F) \text{ ; } \tau \text{ is the graph of } \psi \in C^\infty(\mathbf{T}^n, \mathbf{R}^n),\ \sup_\theta \|D\psi(\theta)\| < \delta_2\}$ and we considered $Y^\infty(F)$ as a subset of $S = \{\psi \in \text{Lip}\,(\mathbf{T}^n, \mathbf{R}^n) \text{ ; } \|\psi\|_{C^0} \leq \delta_1,\ \|D\psi\|_{L^\infty} \leq \delta_2\}$.

On the space S we put the C^0 topology (i.e. uniform convergence), S is then a compact metric space. We denoted by $U = \{F \in \mathcal{S}_n, ,Y^\infty(F) \neq \emptyset\}$. We decompose $Y^\infty(F)$ into $Y^\infty_d(F) = Y^\infty(F) \cap IT^\infty_d(F)$, and $Y^\infty_i(F)$. We define, for $* = i$ or d, $Y_*(F)$ to be the closure of $Y^\infty_*(F)$ in S and $U_* = \{F \in \mathcal{S}_n;\ Y^\infty_*(F) \neq \emptyset\}$. One has $U = U_i \cup U_d$, both the sets U_* are open, and the set valued functions $F \in U_* \to Y^\infty_*(F) \subset S$, and $F \in U_* \to Y_*(F) \subset S$ are lower semi-continuous (this follows from 3.5 and 3.6 using the continuity statement of 3.6, cf. [H₆,12.3]).

7.5. Properties of $\tau \in IT_*(F)$, when $F \in V_*$, for $* = i, d$

- If $\tau \in IT_*(F)$, then τ is non empty, compact, connected, $\tau \subset \mathbf{A}^n_{\delta_1}$ and is invariant by F.
- Since when $\tau \in IT^\infty_*(F)$, τ is homotopic to $\{r = 0\}$, it follows by degree theory, for example, that the map $p_1 \mid \tau : \tau \to \mathbf{T}^n$ is surjective.

Taking limits, we obtain, when $\tau \in IT_*(F)$, $p_1(\tau) = \mathbf{T}^n$.

The same argument gives that, when $\tau \in IT_*(F)$ is included in a tubular neighbourhood N_σ of a C^∞ n-torus σ, homotopic to $\{r = 0\}$ and p_σ denotes the projection of N_σ onto σ, we have

(7.6) $\quad p_\sigma(\tau) = \sigma.$

Caution. If $\tau \in IT(F)$ and $F \in V$, then τ is not always a manifold and its Čech cohomology can be different from the one of $\mathbf{T}^n$ (cf. 5.21).

7.7. If $\tau \in IT_*(F)$, then $F_{|\tau}$ is chain recurrent and one can show that there exists a dense G_δ set $G \subset IT_*(F)$, $F \in V_*$ such that if $\tau \in G$

$F_{|\tau}$ is minimal (every orbit is dense in τ)

and

$F_{|\tau}$ is uniquely ergodic.

Furthermore, using 3.6 and 3.8 (cf. [H_6,6.6]) one can show that $IT_*(F)$, $F \in V_*$, is a perfect topological space (i.e. has no isolated point).

7.8. It follows from 6.4 and 6.9 that $V_i \cap V_d$ (resp. $U_i \cap U_d$) is dense in V_i (resp. U_i) for the C^∞ topology.

PROPOSITION. *We have, if $F \in V_i$ (resp. $F \in U_i$),*

$$IT_d^\infty(F) \cap IT_i(F) = \emptyset$$
$$(\text{resp. } Y_d^\infty(F) \cap Y_i(F) = \emptyset\,.)$$

Proof. Let us prove the first statement, the second being similar. If $IT_d^\infty(F) \cap IT_i(F) \neq \emptyset$, then we can find a sequence $(\tau_j)_{j\in\mathbf{N}} \subset IT_i^\infty(F)$ such that $\tau_j \to \tau$ in the Hausdorff topology. Using 3.8, we can change coordinates and perform the Birkhoff normal form ($k = 1$). In the new coordinates

$$F(\theta, r) = (\theta + \alpha + Br + O(r^2),\ r + O(r^2))\,, \text{ when } r \to 0$$

with a definite matrix B. We denote by $\hat{\tau}_j$ ($j \geq j_0$) the torus τ_j in the new coordinates. We have $\hat{\tau}_j \to \{r = 0\}$ in the Hausdorff topology. By Viterbo's theorem [V], as the torus $\hat{\tau}_j$ is Lagrangian and is homotopic to $\{r = 0\}$, its Maslov class is equal to 0. We now apply [H_3,II.10.6] and conclude that, when $j \geq j_1$, $\hat{\tau}_j$ is the graph of $\psi_j \in C^\infty(\mathbf{T}^n, \mathbf{R}^n)$.

Let $\quad K_j^{-1} \circ F \circ K_j(\theta, 0) = (f_j(\theta), 0)$

with $\quad K_j(\theta, r) = (\theta, r + \psi_j(\theta))$ and

$$D(K_j^{-1} \circ F \circ K_j)(\theta, 0) = \begin{pmatrix} Df_j(\theta) & \bar{b}_j(\theta) \\ 0 & {}^t(Df_j)^{-1}(\theta) \end{pmatrix}.$$

The same argument as in [H_3,I.7], shows that the symmetric matrix $(\bar{b}_j)^{-1}Df_j$ has the same type as B, hence is definite (cf. also [H_6,II.10.2]). We write $f_j = h_j \circ R_{\alpha_j} \circ h_j^{-1}$ where h_j are C^∞ diffeomorphisms of $\mathbf{T}^n$ homotopic to the identity and $\alpha_j \in DC$ (we have $\tau_j \in IT_i^\infty(F)$).

We consider $H_j = T^*h_j$ and $H_j^{-1} \circ K_j^{-1} \circ F \circ K_j \circ H_j$ and calculate the normal torsion on $\{r = 0\}$ and we obtain

$$\int_{\mathbf{T}^n} (Dh_j)^{-1}(Df_j \circ h_j)^{-1}\bar{b}_j \circ h_j\, {}^t(Dh_j)^{-1} d\theta\,.$$

As the integral of definite matrices is definite, we contradict the fact that the normal torsion of F in τ_j is indefinite and that this does not depend on the coordinate system ($F_{|\tau_j}$ is C^∞ conjugated to R_{α_j}, $\alpha_j \in DC$), see 3.4. We arrived to a contradiction and this proves the proposition. $\square$

Caution. The proposition does exclude the possibility that we can find $\tau \in IT_i(F)$ such that $\tau \supset \tau_1$ and $\tau_1 \in IT_d^\infty(F)$.

7.9. Given the set valued function $F \in V_* \to IT^\infty(F) \subset K$ we denote by IT_*^∞ its graph : $IT_*^\infty = \{(F, IT_*^\infty(F)) \subset V_* \times K; F \in V_*\}$

and by $\widehat{IT}_*$ the closure of IT_*^∞ in $V_* \times K$ (on $V_* \times K$ we put the product topology of the C^∞ topology and the Hausdorff topology). The set valued function $F \in V_* \to \widehat{IT}_*(F) = \{\tau \in K\,,\, F \times \tau \in \widehat{IT}_*\}$ is upper semi-continuous and we have $IT_*(F) \subset \widehat{IT}_*(F)$. The elements of the set $\widehat{IT}_*(F)$, $F \in V_*$, are the compact invariant sets obtained by taking limits of sequences $(F_j, \tau_j)_{j \in \mathbf{N}} \subset \widehat{IT}_*$ where $F_j \to F \in V_*$ and $\tau_j \to \tau \in \widehat{IT}_*(F)$, as $j \to +\infty$.

When $F \in V_*$, then the properties of 7.5 hold as well as the fact that $F_{|\tau}$ is chain recurrent on τ.

The examples of 4.23 show that it is possible for $\tau \in \widehat{IT}(F)$ to be homeomorphic to the pseudo-circle $\times \mathbf{T}^{n-1}$ or to $\mathbf{A}_\delta \times \mathbf{T}^{n-1}$, $0 < \delta < \delta_1$, when $n \geq 2$.

7.10

PROPOSITION. *There exists a dense G_δ set for the C^∞ topology, $G \subset V_i$ (resp. $G \subset U_i$) such that if $F \in G$, then*

$$IT_d(F) \supset IT_i(F) \quad (\text{resp.}\, Y_d(F) \supset Y_i(F))\,.$$

Proof. Let us prove the proposition in the first case. What we did in 6.4 and 6.9 shows that, on V_i the closure of IT_d^∞ contains IT_i^∞, hence on V_i, $\widehat{IT}_d \supset \widehat{IT}_i$ and 7.10 follows from [H₃,12.5 prop.1]. □

7.11. We deduce from 7.8 and 7.10 that on a dense G_δ set of V_i, the closed set $IT_d(F)$ contains an open dense set of $IT(F)$. When $\tau \in IT_d^\infty(F)$, $F \in V_d$, using 7.8, [V] and [H₃,II.10.6], in a small neighbourhood of τ in $IT^\infty(F)$ everything reduces, after the Birkhoff normal form (with $k = 1$), to the study of $Y_d^\infty(F)$ with the numbers δ_1 and δ_2 of 7.2, depending on τ.

We refer the reader to [H₆,8] for generic dynamical properties that can occur on a generic torus. We would like to add that there exists a dense G_δ set $G \subset IT_d(F)$ such that if $\tau \in G$, then τ is C^1 Lagrangian torus of $\mathbf{A}^n$ (cf. [H₆,7.8]).

7.12. Using 7.6, 5.2, 3.8 and [H₆,7.8] one can prove the following proposition.

PROPOSITION. *There exists a dense G_δ set for the C^∞ topology, $G_2 \subset V_i$ such that if $F \in G_2$, then there exists an open dense set $W \subset IT_i(F)$ with the property, if $\tau \in W$, then τ is not (stably) a graph on $\mathbf{T}^n \times \{0\}$: $p_{1|\tau} : \tau \to \mathbf{T}^n$ is not injective.*

REFERENCES

[A] D.V. Anosov and A.B. Katok, *New examples in smooth ergodic theory*, Trans Moscow Math. Soc. **23** (1970), pp. 1–35.

[B_1] J.K. Beem and P.E. Ehrlich, *Geodesic completeness and stability*, Math. Proc. Camb. Phil. Soc. **102** (1987), pp. 319–328.

[B_2] G.D. Birkhoff, *Note sur la stabilité*, **15** (1936), pp. 339–344; Collected papers, Vol. 2, pp. 662–667.

[B_3] J. Bost, *Tores invariants des systèmes dynamiques hamiltoniens*, Séminaire Bourbaki, exposé n° 639, Astérisque **133-134** (1986), pp. 113–157.

[D_1] R. Douady, *Stabilité ou instabilité des points fixes elliptiques*, Ann. Scient. Ec. Norm. Sup., 4ème série **21** (1988), pp. 1–46.

[D_2] R. Douady, *Applications du théorème des tores invariants*, Thèse Univ. Paris VII (1982), Chap I ; announced in, *Une démonstration directe de l'équivalence des théorèmes de tores invariants pour les diffeomorphisms et les champs de vecteurs*, C.R. Acad. Sci. Paris **295** (1982), pp. 201–204.

[F] A. Fathi and M.R. Herman, *Existence de difféomorphismes minimaux*, in Dynamical Systems I - Warsaw, Astérisque **49** (1977), pp. 37–59.

[G_1] W.H. Gottschalk and G.A. Hedlund, *Topological dynamic*, Amer. Math. Soc. (1955) §14.

[G_2] D. Gromoll and W. Meyer, *On differentiable functions with isolated critical points*, Topology **8** (1969), pp. 361–369.

[H_1] M. Handel, *A pathological area preserving C^∞ diffeomorphism of the plane*, Proc. Amer. Math. Soc. **86** (1982), pp. 163–168.

[H_2] MM. Hupaev, *Generalization of Ljapunov's second method and the investigation of some resonance problems*, Soviet. Math. Dokl. **11** (1970), pp. 868–872.

[H_3] M.R. Herman, *Inégalités a priori pour des tores lagrangiens invariants par des difféomorphismes symplectiques*, Publ. Math. I.H.E.S. **70** (1990), pp. 47–101.

[H_4] M.R. Herman, *Sur la conjugaison différentiable du cercle à des rotations*, Publ. Math. I.H.E.S. **49** (1979), pp. 5–233.

[H_5] M.R. Herman, *Sur les courbes invariantes par les difféomorphismes de l'anneau*, Vol.1, Astérisque **103-104** (1983).

[H_6] M.R. Herman, *On the dynamics on Lagrangian invariant tori by symplectic diffeomorphisms*, to appear in Proc. Conf. Aquilla 1990, *On variational methods in Hamiltonian systems and elliptic equations.*

[H_7] M.R. Herman, *Construction de difféomorphismes ergodiques*, Manuscript, 1979.

[H_8] M.R. Herman, *Existence et non existence de tores invariants par des difféomorphismes symplectiques*, Séminaire E.D.P., Publ. Centre de Math. Ecole Polytechnique (1987-1988), Exposé XIV.

[H_9] N.J. Hicks, *Notes on differential geometry*, Van Nostrand, 1965.

[K] A.B. Krygin, *Examples of ergodic cylindrical cascades*, Math. Notes **16** (1974), pp. 1180–1186.

[M_1] D. Montgomery and L. Zippin, *Topological transformation groups*, J. Wiley, 1955.

[M_2] J. Moser, *On the elimination of the irrationability condition and Birkhoff's concept of complete stability*, Bol. Soc. Mat. Mexicano **5** (1960), pp. 167–175.

[N] N.N. Nokoroshev, *An exponential estimate of the time stability of nearly-integrable Hamiltonian systems*, Russian Math. Surveys **32**, 6 (1977), pp. 1–65.

[P] J. Palis, Jr. and W. de Melo, *Geometric theory of dynamical system*, Springer Verlag, New York, (1982).

[V] C. Viterbo, *New obstruction to embedding Lagrangian tori*, Inventiones Math. **100** (1990), pp. 301–320.

[Y] J.C. Yoccoz, *Conjugaison différentiable des difféomorphismes du cercle dont le nombre de rotation vérifie une condition diophantienne*, Ann. Scient. Ec. Norm. Sup., 4ème série **17** (1984), pp. 333–359.

MINIMAL ORBITS FOR SMALL PERTURBATIONS OF COMPLETELY INTEGRABLE HAMILTONIAN SYSTEMS

A. KATOK*

1. Introduction. In this paper we summarize an attempt to carry out certain aspects of Aubry–Mather theory for twist maps (see e.g. [M1] [AL] [K] [Mos2] [B1]) to Hamiltonian systems with more than two degrees of freedom. In a sense, the paper should be considered as a continuation of [BK]. Although the minimal motions for an arbitrary "admissible" rotation vector, may not exist (see [He2], p. 54) the combination of KAM theory with some of the methods developed in [BK] still yields a considerable information about those motions. Our two main results are:

(i) the orbits in KAM tori are minimal and they are the only minimal motions for corresponding rotation vectors (Section 5, Theorem 1);

(ii) there are infinitely many rotation vectors for which KAM tori do not exist but minimal motions do exist and the closure of these minimal motions contains all the KAM tori (Section 6, Theorem 2).

Both results hold for small perturbations of completely integrable Hamiltonian systems satisfying convexity assumptions (see Section 2 below and [BK], Sections 2 and 7). For (ii) we naturally assume that the set of KAM tori is nowhere dense.

John Mather has a proof of (i) or a close result (personal communication) which looks considerably more involved that the proof presented in this paper.

We will use notations from [BK]. We will only present arguments for the discrete-time case, namely for symplectic maps. The reduction of the more traditional continuous-time case of Hamiltonian vector-fields to the discrete-time one is explained in Section 7 of [BK]. Furthermore, some of the arguments and estimates presented in [BK] for Birkhoff periodic orbits work almost literally for more general types of minimal orbits and orbit segments considered in this paper. In such cases we will give precise formulations and will refer to an appropriate place in [BK] for elaboration.

This paper grew out of the talk given on September 17, 1988 at the Conference on Hamiltonian Dynamics at La Jolla, California dedicated to John Greene's sixtieth birthday and was certainly influenced by Greene's work. I would like to thank Robert McKay for inviting me to speak at the conference. Earlier discussions with Victor Bangert were useful in developing some of the ideas which led to this paper.

---

*Mathematics 253–37, California Institute of Technology, Pasadena, CA 91125.
Current address: Department of Mathematics, Pennsylvania State University, University Park, PA 16802.

2. Minimal states. Let us briefly recall the notations and assumptions from [BK], Section 2. We will consider the space

$$M = \mathbf{T}^n \times \mathbf{R}^n = \{\varphi_1, \ldots, \varphi_n, r_1, \ldots, r_n); \quad \varphi_i \in \mathbf{R}/\mathbf{T}, r_i \in \mathbf{R}\}$$

provided with the standard symplectic form $\Omega = \sum_{i=1}^{n} d\varphi_i \wedge dr_i$ and a symplectic diffeomorphic embedding

$$f : \mathbf{T}^n \times U \to M$$

where $U \subset \mathbf{R}^n$ is diffeomorphic to an open n-disc. We assume that f is a small perturbation of an integrable map $f_0 : f_0(\varphi, r) = (\varphi + \alpha(r), r)$. Let us denote by F and F_0 the lifts of f and f_0 correspondingly to the universal cover of $\mathbf{T}^n \times U$. Let $F_0(x, r) = (x + a(r), r)$. In terms of generating functions, F_0 is generated by a function $h(x' - x)$ and F by a function

$$H(x, x') = h(x' - x) + P(x, x')$$

where P is periodic, i.e.; $P(x + m, x' + m) = P(x, x')$ for all $m \in \mathbf{T}^n$, and is small with several derivatives. Precise assumptions on the size of P will vary. The strongest assumptions will be those to guarantee that the map f has sufficiently many invariant KAM tori close to the tori $r = \text{const.}$ for the unperturbed map f_0.

On the other hand, we will assume that the function h is strictly differentiably convex, i.e. its Hessian is a positive definite quadratic form.

At the end of Section 2 of [BK] an extension of H to the whole space $\mathbf{R}^n \times \mathbf{R}^n$ is described which allows to keep the smallness of the perturbation part P of the generating function. We will use that extension, but unlike [BK] we will still denote the extended function by $H = h + P$.

Fix $x, y \in \mathbf{R}^n$ and a natural number q. Let

$$\Psi_q^{x,y} = \{x_0 = x, x_1 \ldots x_{q-1}, x_q = y;\ x_1, \ldots x_{q-1} \in \mathbf{R}^n\}.$$

Let us introduce the Lagrangian $L_q^{x,y} : \Psi_q^{x,y} \to \mathbf{R}$ by

$$L_q^{x,y}(x_0, x_1, \ldots, x_q) = \sum_{i=0}^{q-1} H(x_i, x_{i+1}).$$

Any critical point $(x_0, x_1, \ldots, x_q)$ of the Lagrangian determines a unique orbit segment of F such that $(x_i, r_i) = F^i(x_0, r_0)$ and vice versa, for any such orbit segment the sequence of first coordinates $(x, x_1, \ldots, x_{q-1}, y)$ is a critical point for $L_q^{x,y}$. Sometimes we will call the elements of the spaces $\Psi_q^{x,y}$ *states* and the critical points of $L_q^{x,y}$ *equilibrium states*.

The convexity of h and the smallness of P imply that $L_q^{x,y}$ is a proper function bounded from below and hence that it reaches its absolute minimum which we will denote by $\ell_q^{x,y}$.

DEFINITION. Any state $(x, x_1, \ldots, x_{q-1}, y) \in \Psi_q^{x,y}$ for which $L_q^{x,y}(x, x_1 \ldots, x_{q-1}, y) = \ell_q^{x,y}$ is called a *minimal state*. Corresponding orbit segments for F and f are called *minimal orbit segments*.

Let δ_k be the C^k norm of the function P. For a state $\overline{x} \in \Psi_q^{x,y}$ let us denote $x_i - x_{i-1} = a_i$. The following statements are direct counterparts of corresponding results from [BK] for minimal periodic orbits. The letter C with various indices denotes constants which depend only on the unperturbed generating function h, i.e. on the map f_0.

LEMMA 1. *If $\overline{x} \in \Psi_q^{x,y}$ is an equilibrium state then $|a_{i+1} - a_i| < C_1\delta_1$ for $i = 1, \ldots, q-1$. If $\overline{x}$ is a minimal state then $|a_{i+1} - a_i| < C_2\delta_0^{1/2}$.*

Proof. See [BK], Lemma 1.

LEMMA 2. *Let $\overline{x}$ be a minimal state for $L_q^{x,y}$ and $q > C_3\delta_1^{-1/2}$. Then $|a_i - a_j| < C_4\delta_1^{1/2}$.*

Similarly, if $q > C_5\delta_0^{-1/4}$ then $|a_i - a_j| < C_6\delta_0^{1/4}$.

Proof. See the proof of Lemma 2 from [BK]. It works verbatim in our case if either $C\delta_1^{-1/2} < \min(i,j)$ or $\max(i,j) < q - C\delta_1^{-1/2}$ for the first statement and either $C\delta_0^{-1/4} < \min(i,j)$ or $\max(i,j) < q - C\delta_0^{-1/4}$ for the second. If both conditions are violated let $k = \left[\frac{q}{2}\right]$ and then our inequality for q guarantees that one of the conditions holds for the pair (i,k) and the other for the pair (j,k). Since $|a_i - a_j| \leq |a_i - a_k| + |a_k - a_j|$, by doubling the constant we obtain the desired inequalities for arbitrary i and j.

□

Lemma 2 implies that if the vector $v = \dfrac{y-x}{q} \in a(U)$ and is not too close to the boundary of $a(U)$ then all minimal orbit segments for f corresponding the minimal states from $\Psi_q^{x,y}$ belong to $\mathbf{T}^n \times U$ and consequently they are orbit segments of the original map. Thus the way we extended the generating function is unimportant.

LEMMA 3. *Let $(\varphi_i, r_i) = f^i(\varphi_0, r_0)$, $i = 0, 1, \ldots q$ be a minimal orbit segment, $i, j, k \in \{0, 1, \ldots q\}$ be different. Then*

$$|r_i - r_k| < C_7(\operatorname{dist}(\varphi_i, \varphi_j) + \operatorname{dist}(\varphi_i, \varphi_k))^{1/2}$$

Proof. See [BK], Proposition 5.

3. Minimal orbits

DEFINITION. An orbit of f is *minimal* if every finite segment of it is a minimal orbit segment.

For the geodesic problem on a Riemannian manifold, the corresponding concept of minimal geodesic was considered by Morse [Mo], Hedlund [H] and recently by Bangert [B2].

DEFINITION. An orbit of f has *rotation vector* v if for some (and hence any) lift $(x_m, r_m) = F^m(x_0, r_0)$, $m \in \mathbf{Z}$

$$\lim_{m \to \pm\infty} \frac{x_m - x_0}{m} = v. \tag{1}$$

More generally, the *rotation set* of the orbit, is the set of all limit points of $\frac{x_m - x_0}{m}$. For $(\varphi, r) \in \mathbf{T}^n \times U$ let $\rho(\varphi, r)$ be the vector opposite to the difference between the x coordinates of any lift of (φ, r) and its F-image.

DEFINITION. For any probability Borel f-invariant measure μ the *rotation vector* of μ, $\rho(\mu)$ is

$$\rho(\mu) = \int_{\mathbf{T}^n \times U} \rho(\varphi, r) d\mu$$

By Birkhoff Ergodic Theorem for any finite f-invariant measure μ, μ-almost every orbit has a rotation vector i.e. the limit in the left-hand part of (1) exists. In particular, if μ is ergodic, μ-almost every orbit of f has the same rotation vector equal to $\rho(\mu)$.

PROPOSITION 1. *Let $(\varphi_m, r_m) = f^m(\varphi_0, r_0)$, $m \in \mathbf{Z}$ be a minimal orbit of f.*

If $\varphi = \varphi_k$ is a non-isolated point for the set $\mathbf{\Phi} = \{\varphi_i\}_i \in \mathbf{Z}$ then for every integer i

$$|r_i - r_k| \le C_8(\text{dist}(\varphi_i, \varphi))^{1/2} \tag{2}$$

If the orbit is non-periodic and φ is an isolated point in $\mathbf{\Phi}$, then (2) holds for all the integers i, except maybe for one.

If the orbit is periodic with the minimal period q then (2) holds for all except may be one $i \in \{0, 1, \ldots, q-1\}$.

Proof. If φ is a non-isolated point in $\mathbf{\Phi}$ then for any i one can choose an integer j such that $\text{dist}(\varphi, \varphi_i) < \text{dist}(\varphi, \varphi_i)$ and (2) immediately follows from Lemma 3.

Otherwise choose j such that dist $(\varphi, \varphi_j) < 2 \inf_{i \in \mathbf{Z}} \text{dist}(\varphi, \varphi_i)$ and again apply Lemma 3.

□

PROPOSITION 2. *If a minimal orbit has rotation vector v then for all integers n, $|r_n - a^{-1}(v)| < C_9 \delta_1^{1/2}$.*

Proof. Consider the lift of the orbit and the corresponding minimal state $(\ldots x_{-1}, x_0, x_1 \ldots)$. Take a sufficiently large $N > |n|$ such that

$$\left| \frac{x_N - x_{-N}}{2N+1} - v \right| < \delta_1^{1/2}. \tag{3}$$

Since $\overline{x} = (x_{-N'} \ldots, x_N) \in \Psi_{2N}^{x_{-N'} x_N}$ is a minimal state, by Lemma 2 for any $i, j \in \{-N+1, \ldots, N\}$ $|a_i - a_j| < C_4 \delta_1^{1/2}$ where as before $a_i = x_i - x_{i-1}$.

We have $\sum_{i=N+1}^{N} a_i = x_N - x_{-N}$, hence by (3) for any i

$$|a_i - v| < (C_4 + 1)\delta_1^{1/2}. \tag{4}$$

But since $\overline{x}$ is an equilibrium state we have

$$r_i = \frac{\partial H}{\partial x}(x_i, x_{i+1}) = dh(a_i) + \frac{\partial P}{\partial x}(x_i, x_{i+1}), \tag{5}$$

where $\left\|\frac{\partial P}{\partial x}\right\| \leq \delta_1$ by the definition of δ_1, and $dh(v) = a^{-1}(v)$. Thus by (4) and (5) $|r_n - a^{-1}(v)| < \left\|\frac{\partial P}{\partial x}\right\| + |dh(a_i) - a^{-1}(v)| \leq \delta_1 + |dh(a_i) - dh(v)| \leq \delta_1 + \|dh\|\,(C_4 + 1)\delta_1^{1/2}$.

□

The next statement is an immediate corollary of the definitions of a minimal orbit segment and a minimal orbit. We formulate it separately because of its importance for the theory.

PROPOSITION 3. *Let $x \in \mathbf{R}^n, x = \lim_{m\to\infty} x_m$ and suppose there are $k_m \to \infty, l_m \to \infty$ and minimal states $\overline{x}^{(m)} = (x_0^{(m)}, \ldots, x_{k_m+l_m}^{(m)}) \in \Psi_{k_m+l_m}^{x_0^{(m)}, x_{k_m+l_m}^{(m)}}$ such that $x_m = x_{k_m}^{(m)}$. Then the F-orbit of x is minimal. In particular, the closure of a minimal f-orbit consists of minimal f-orbits.*

Let us fix a vector $v \in a(U)$ not too close to the boundary of $a(U)$. Consider a sequence of minimal states in $\Psi_{2q_m}^{x_m, x_m + 2q_m v}$ where x_m and $q_m \to \infty$ are chosen in such a way that the sequence $x_m + q_m v$ converges. This is always possible because x_m can be moved by any integer vector so we can assume that the vectors $x_m + q_m v$ lie within the unit cell and then take a converging subsequence. Put a uniform normalized δ-measure on the projection of the minimal orbit segment determined by each minimal state and take a weakly converging subsequence. The limit measure is f-invariant and has rotation vector v. It is also supported on the set of minimal orbits since for every point $(\varphi, r) \in \operatorname{supp}\mu$ the x-coordinate of every lift of φ satisfies the assumption of Proposition 3. If μ also happens to be ergodic, then μ-almost every orbit has rotation vector v.

4. Minimal action function

LEMMA 4. *For any $x, y, w \in \mathbf{R}^n$ and a natural number q*

$$|\ell_q^{x,x+w} - \ell_q^{y,y+w}| \leq C \tag{6}$$

where C is independent of x, y, w, q.

Proof. First, let us notice that for every $p \in \mathbf{Z}^n$, $\ell_q^{x+p,\, x+w+p} = \ell_q^{x,\, x+w}$. Then let us find $p \in \mathbf{Z}^n$ such that the distance between $x+p$ and y is less than $n^{1/2}$. Then compare the values of the Lagrangian at any minimal state $(y, y_1, \dots, y_{q-1}, y+w) \in \Psi_q^{y,\, y+w}$ and at the state $(x+p, y_1, \dots, y_{q-1}, x+p+w) \in \Psi_q^{x+p,\, x+p+2}$. The first value is equal to $\ell_q^{y,\, y+w}$ by definition, the second is greater or equal than $\ell_q^{x,\, x+w}$. Thus we have $\ell_q^{x,\, x+w} \le \ell_q^{y, y+w} + |H(x+p, y_1) - H(y, y_1)| + |H(y_{q-1},\, x+p+w) - H(y_{q-1},\, y+w)| \le \ell_q^{y,\, y+w} + |h(x+p-y_1) - h(y-y_1)| + |h(x+p+w-y_{q-1}) - h(y+w-y_{q-1})| + |P(x+p,\, y_1) - P(y,\, y_1)| + |P(y_{q-1}, x+p+w) - P(y_{q-1}, y+w)| \le \ell_q^{y, y+w} + 2|h(x+p-y)| + 2\delta_0$. To obtain the last inequality we used the convexity of the (extended) function h. Since $\|x+p-y\| < n^{1/2}$, the lemma is proved.

□

PROPOSITION 4. *The ratio* $\dfrac{\ell_q^{x,\, x+qv}}{q}$ *has a limit as* $q \to \infty$, *which is independent of* x.

Proof. By the definition of the minimal state one has

$$\ell_{q_1+q_2}^{x, x+(q_1+q_2)v} \le \ell_{q_2}^{x, x+q_1 v} + \ell_{q_2}^{x+q_1 v, x+(q_1+q_2)v} \tag{7}$$

Combining (6) and (7) one derives "almost subadditivity" of $\ell_q^{x, x+qv}$ in q

$$\ell_{q_1+q_2}^{x, x+(q_1+q_2)v} \le \ell_{q_1}^{x, x+q_1 v} + \ell_{q_2}^{x, x+q_2 v} + C$$

which implies the existence of $\lim\limits_{q\to\infty} \dfrac{\ell_q^{x, x+qv}}{q}$. Lemma 4 implies that the limit does not depend on x.

□

We will denote $\lim\limits_{q\to\infty} \dfrac{\ell_q^{x, x+qv}}{q}$ by $\mathcal{L}(v)$ and will call $\mathcal{L}$ the *minimal action function*.

PROPOSITION 5. *$\mathcal{L}$ is a convex function.*

Proof. Using the definition of the minimal state and Lemma 4 we have

$$\ell_{2q}^{x, x+v+w} \le \ell_q^{x, x+v} + \ell_q^{x+v, x+v+w} \le \ell_q^{x, x+v} + \ell_q^{x, x+w} + C$$

which immediately implies that $\mathcal{L}\left(\dfrac{v+w}{2}\right) \le \dfrac{\mathcal{L}(v) + \mathcal{L}(w)}{2}$.

□

Let us fix a lift of $\varphi \in \mathbf{T}^n$ and denote it as usual by x; let furthermore $x'(x, r)$ be the first coordinate of $F(x, r)$. The periodicity of H implies that the function $H(x, x'(x, r))$ does not depend on the choice of a particular lift of φ and thus can be denoted by $\mathcal{H}(\varphi, r)$.

PROPOSITION 6. $\mathcal{L}(v) = \inf \int \mathcal{H} d\mu$ *where the infinum is taken over all probability f-invariant measures μ with rotation vector v.*

Proof. (1) In order to establish the inequality

$$\mathcal{L}(v) \le \int \mathcal{H} d\mu \tag{8}$$

for every μ with rotation vector v it is enough to do that only for ergodic measures μ. For, otherwise take the ergodic decomposition of μ,

$$\mu = \int_Z \mu_z d\nu.$$

We have $v = \rho(\mu) = \int_Z \rho(\mu_z) d\nu$, since the rotation vector is defined as an integral of the shift function $\rho(\varphi, r)$.

On the other hand, also by definition

$$\int \mathcal{H} d\mu = \int_Z \left(\int \mathcal{H} d\mu_z \right) d\nu.$$

Assuming that $\mathcal{L}(\rho(\mu_z)) \leq \int \mathcal{H} d\mu_z$ and using the convexity of $\mathcal{L}$: $\mathcal{L}(v) \leq \int_Z \mathcal{L}(\rho(\mu_z)) d\nu$ we obtain (8).

For an ergodic measure μ with $\rho(\mu) = v$, as we mentioned already, μ-almost every point (φ, r) has rotation vector v. That means that for every lift (x, r) of (φ, r), the x-coordinate of $F^q(x, r)$ has the form

$$x + w_q(x, r) = x + qv + o(q).$$

Thus

$$\ell_q^{x, x+w_q(x,r)} \leq \sum_{i=0}^{q-1} \mathcal{H}(f^i(\varphi, r)) = q \int \mathcal{H} d\mu + o(q).$$

Now, let us take a minimal state $\overline{x} \in \Psi_q^{x, x+w_q(x,r)}$ and use Lemma 2 which allows to replace its last $C\delta_1^{-1/2}$ $o(q)$ terms to obtain a state $\overline{\overline{x}} \in \Psi_q^{x, x+qv}$ such that the norm of the difference of corresponding coordinates for $\overline{x}$ and $\overline{\overline{x}}$ does not exceed $C'\delta^{1/2}$. This implies that $\ell_q^{x,x+qv} \leq q\mathcal{L}(\overline{\overline{x}}) \leq \ell_q^{x,x+w_q(x,r)} + o(q) \leq q \int \mathcal{H} d\mu + o(q)$. Dividing by q and taking limit we obtain (8).

(2) In order to construct a measure μ with rotation vector v (unfortunately, not necessarily ergodic) for which $\mathcal{L}(v) = \int \mathcal{H} d\mu$ we take a sequence of absolute minima in $\Psi_{2q}^{x-qv,\, x+qv}$ and denote by μ_q the normalized δ-measures on corresponding minimal orbit segments for f. Since for any q

$$\frac{1}{2q}\, \ell_{2q}^{x-qv, x+qv} = \int \mathcal{H}\, d\mu_q,$$

for any weak limit point μ of the sequence μ_q, $\mathcal{L}(v) = \int \mathcal{H}\, d\mu$.

Proposition 7. *If v is an extreme point of the convex set $\mathcal{L}_v = \{w : \mathcal{L}(w) \leq \mathcal{L}(v)\}$ then there exists an ergodic measure μ with rotation vector v for which*

$$\mathcal{L}(v) = \int \mathcal{H} d\mu$$

and whose support consists of minimal orbits.

Proof. Consider the construction from part (2) of the previous proposition. The support of the limit measure μ consists of orbits which are limits of minimal orbit segments of increasing length. In order to see that one should remember that any subsegment of a minimal orbit segment is minimal. If the measure μ is ergodic, it satisfies the assertion of the proposition.

Otherwise consider the ergodic decomposition of $\mu, \mu = \int_Z \mu_z d\nu$, so that $v = \int_Z \rho(\mu_z)d\nu$ and by Proposition 6

$$\mathcal{L}(v) = \int \mathcal{H}d\mu = \int_Z \left(\int \mathcal{H}d\mu_z\right) d\nu \geq \int_Z \ell(\rho(u_z))d\nu \tag{9}$$

Since $\mathcal{L}$ is convex and v is an extreme point of the set $\mathcal{L}_v$ then either

$$\mathcal{L}(v) < \int_Z \ell(\rho(\mu_z))d\nu$$

which contradicts (9) or for ν-almost every $z \in Z$, $\rho(\mu_z) = v$ and every such measure μ_z satisfies the assertion of the proposition.

□

5. Minimality of KAM tori. Let us summarize without going into much technical details some of the results concerning the existence of invariant tori which are needed for the subsequent discussion (see e.g. [Bo]; the original proof for the real analytic case is in [A] and useful discussions are in [Mos1] and [He1]):

There exists a closed set $V \subset a(U)$ of a large relative Lebesque measure such that for any map f for which the perturbation part P of the generating function H is small enough with sufficiently many derivatives and for any $v \in V$, there is an f-invariant torus

$$T_{f,v} = \{(\varphi, r) : r = g_{f,v}(\varphi)\}$$

such that the restriction of f to $T_{f,v}$ has rotation vector v. Those tori are usually called KAM tori. The set V is defined by arithmetic conditions which have to do with rational approximation. Let us point out two important uniformity properties of KAM tori which will play an important part in our proofs of minimality and uniqueness.

First, the difference $g_{f,v}(\varphi) - a^{-1}(v)$ is uniformly small with several derivatives.

Secondly, the restriction of the map f to the torus $T_{f,v}$ is smoothly conjugate to the translation $L_v : \varphi \to \varphi + v$ via a diffeomorphism whose φ-coordinate $\psi_{f,v} : \mathbf{T}^n \to \mathbf{T}^n$, is uniformly in v and f close to the identity with several derivatives.

Not every invariant torus of the form graph g where $g : \mathbf{T}^n \to U$ which the map f may happen to possess is necessarily smooth and even if it is, it might not satisfy the uniform estimates mentioned above. Such extra or "accidental" tori are not covered by our results concerning minimality and uniqueness although they satisfy local versions of some of those properties.

Let us fix a KAM torus $T_{f,v}$ and consider the following symplectic coordinate change $S^{f,v} = S_1^{f,v} \circ S_2^{f,v}$, where

$$S_1^{f,v}(\varphi, r) = (\varphi, r - g_{f,v}(\varphi)), \qquad S_2^{f,v}(\varphi, r) = (\psi_{f,v}(\varphi), (d\psi^*_{f,v})_\varphi^{-1} r).$$

Here $(d\psi^*_{f,v})\varphi$ denotes the matrix transposed to that of the derivative of $\psi_{f,v}$ at φ. The map $S_1^{f,v}$ is symplectic because the torus $T_{f,v}$ is a Lagrangian manifold. From now on let us suppress the dependence on f and v in our notation.

The coordinate change S transforms our KAM torus $T_{f,v} = T$ into the standard torus $r = 0$ and the map f restricted to T, into the linear translation L_v. It is important to remember that the estimates given by Lemmas 1-3 and results based on those lemmas remain true, maybe with different constants, due to the uniformity of the functions $g_{f,v}$ and maps $\psi_{f,v}$.

In general, generating functions change under symplectic coordinate changes in a complicated way. However the map S_2 is "Lagrangian" and it carries out the generating function. The lift of the map g to $\mathbf{R}^n$ has the form dG where $G : \mathbf{R}^n \to \mathbf{R}$. It is easy to see that the map S_1 changes the generating function by adding a coboundary $G(x') - G(x)$. Thus, minimal orbit segments and minimal orbits are preserved under the coordinate change S. S_2 does not change the minimal action function $\mathcal{L}$ either; S_1 may only add a linear term to it and thus does not change properties like strict convexity, etc.

After the coordinate change the generating function takes the form

$$Q(x' - x - v) + P(x, x')$$

where Q is a positive definite quadratic form and both the first and the second derivatives of P vanish at $x' - x = v$. By adding a constant we may assume that P itself vanishes too.

Thus, in a fixed neighborhood of the plane $x' - x = v$

(10) $$P(x, x') < CQ(x' - x - v)^{3/2} \leq \frac{1}{2}\, Q(x' - x - v)$$

Since derivatives of the maps S and S^{-1} are uniformly bounded, that neighborhood contains pairs (x_i, x_{i+1}) for any minimal state from $\psi_q^{x,x+qv}$. We will use our new coordinates to calculate those minimal states.

LEMMA 5. *The only minimal state in* $\psi_q^{x,x+qv}$ *is* $(x, x+v, \ldots, x+qv)$.

Proof. Let $(x = x_0, x_1, \ldots, x_{q-1},\ x + qv = x_q)$ be a minimal stade. Then, using (10) we have

$$L_1^{x,x+qv}(x_0, x_1, \ldots, x_{q-1}, x_q) = \sum_{i=0}^{q-1} Q(x_{i+1} - x_i - v) + P(x_i, x_{i+1}) \geq$$

$$\frac{1}{2}\sum_{i=0}^{q-1} Q(x_{i+1} - x_i - v) > 0 \text{ unless } x_{i+1} = x_i + v \text{ i.e. } x_i = x_0 + iv,\ i = 0, 1, \ldots, q.$$

□

COROLLARY 1. *Any KAM torus consists of minimal orbits.*

LEMMA 6. *There exists $\varepsilon > 0$ which depends only on the unperturbed map f_0 and on the C^k size of the perturbation for some k, such that if for a vector v there is a KAM torus with rotation vector v, than for $\|v' - v\| < \varepsilon$*

$$\mathcal{L}(v) + \ell_v(v'-v) + Q_v(v'-v) - C\|v'-v\|^3 \leq \mathcal{L}(v') \leq \mathcal{L}(v) + \ell_v(v'-v) + Q_v(v'-v) + C\|v'-v\|^3,$$

where ℓ_v is a linear function and Q_v is a positive definite quadratic form.

Proof. The coordinate changes described at the beginning of this section may only change $\mathcal{L}$ by a linear term. Thus we will use the new coordinates for our calculations.

Let $\overline{x} \in \psi_q^{x,x+qv'}$ be a minimal state.

We have

$$qQ(v'-v) + Cq\|v'-v\|^3 \geq qQ(v'-v) + \sum_{i=0}^{q-1} P(x+(i+1)v', x+iv') =$$
$$L_q^{x,x+qv'}(x, x+v', \ldots, x+(q-1)v', x+qv') \geq \ell_q^{x,x+qv'} =$$
$$L_q^{x,x+qv'}(x_0, x_1, \ldots, x_{q-1}, x_q) = \sum_{i=0}^{q-1} Q(x_{i+1} - x_i - v) + P(x_{i+1}, x_i).$$

Using (10) and the convexity of Q we can continue

$$\sum_{i=0}^{q-1} Q(x_{i+1} - x_i - v) + P(x_i, x_{i+1}) \geq \sum_{i=0}^{q-1} Q(x_{i+1} - x_i - v) - C\big(Q(x_{i+1} - x_i - v)\big)^{3/2} \geq$$
$$q(Q(v'-v) - CQ(v'-v)^{3/2}) \geq qQ(v'-v) - C'q\|v'-v\|^3.$$

□

THEOREM 1. *If for a vector v there exists a KAM torus with that rotation vector then every minimal orbit with rotation vector v (or even those whose rotation set contains v) belongs to the torus.*

Proof. Take the closure of such a minimal orbit. It is an f-invariant closed set. Its intersection with the torus is also f-invariant and closed. Since every orbit of f on the torus is dense, the intersection is either empty or coincides with the torus. By Proposition 1 in the latter case there is no room for elements of a minimal orbit outside the torus; hence the whole orbit lies on the torus.

Consider the former case. Take a lift of the orbit, $F^m(x,r) = (x_m, r_m), m \in \mathbf{Z}$. By our assumption there exists either a sequence $q_m \to \infty$ or $q_m \to -\infty$ such that $x_{q_m} = x + q_m v + o(q_m)$. The two cases are completely symmetric so we assume the first.

Since the orbit closure does not intersect the torus,
$Q(x_{i+1}-x_i-v)\geq\delta>0$ for all i. Thus, using (10) we obtain

$$L_{q_m}^{x,x_{q_m}}(x,x_1,\dots,x_{q_m})=\sum_{i=0}^{q_m-1}Q(x_{i+1}-x_i-v)+P(x_i,x_{i+1})\geq\frac{1}{2}\sum_{i=0}^{q_m-1}Q(x_{i+1}-x_i-v)>\frac{q_m\delta}{2}.$$

On the other hand, let $u_m=\dfrac{x_{q_m}-x}{q_m}$ so that $u_m=v+o(1)$ and let us calculate and estimate using (10) again

$$L_{q_m}^{x,x_{q_m}}(x,x+u_m,x+2u_m,\dots,x+(q_m-1)u_m,x_{q_m})=\sum_{i=0}^{q_m-1}Q(u_m)+P(x+iu_m,x+$$
$$(i+1)u_m)\leq\frac{3}{2}q_mQ(u_m).$$

Since for large enough m, $Q(u_m)<\dfrac{\delta}{3}$, $(x,x_1,\dots,x_{q_m})$ is not a minimal orbit segment so our orbit is not minimal.

□

6. Other minimal orbits. Now, having established the minimality of KAM tori and their rather strong uniqueness properties we are going to look for minimal orbits which are not associated with these tori. Let W be the set of all vectors for which KAM tori do not exist and let $W_M\subset W$ be the set all $w\in W$ for which a minimal orbit with the rotation vector w exists. Let A_K be the set of rotation vectors for which a KAM torus exists. If follows from Proposition 7 that the convex hull of $A_M=A_K\cup W_M$ contains W. Now we are going to strengthen that statement.

PROPOSITION 8. *The convex hull of W_M contains W.*

Proof. By definition A_M is the set of all vectors v for which minimal orbits with rotation number v exists. By the convexity of the minimal action function $\mathcal{L}$ every w is a linear combination of extreme points of the set $\mathcal{L}_w$ which by Proposition 7 belong to A_M. What remains to prove is that for $w\in W\setminus W_M$ none of those extreme ponts can belong to the set A_K, i.e. correspond to a KAM torus. But that follows from Lemma 6, since after a coordinate change which only changes $\mathcal{L}$ by a linear term any vector $v\in A_K$ becomes an isolated minimum of $\mathcal{L}$.

□

THEOREM 2. *Unless KAM tori exist for every v, there are infinitely many v for which a KAM torus does not exist but minimal orbits with rotation number v exist. Furthermore, the closure of the set of such v contains the boundary of the set $a(U)\setminus W$.*

Proof. Follows immediately from Proposition 8.

The most natural approach to the construction of minimal orbits is the one indicated in part (2), Proposition 6 and further discussed in the proof of Proposition 7. In other words, one should hope that in general minimal orbit segments with correct average behavior converge to a minimal orbit with the desirable rotation vector. The kind of convergence discussed in our proof of Propositions 6 and 7 is the weak convergence of δ-measures on minimal orbit segments. Another type of convergence would be the convergence in Hausdorff topology on the space of compact subsets of $\mathbf{T}^n \times U$. In order to guarantee the convergence one has to choose the initial condition carefully. A natural way to do that is to start from minimal Birkhoff periodic orbits constructed in [BK] (see Corollaries 3 and 4 of that paper). Any segment of such an orbit, whose length does not exceed the period, is a minimal orbit segment in our sense.

Now we will give a simple criterion for the existence of minimal orbits with a given rotation number whose validity is supported by rather convincing numerical experiments performed by Mark Muldoon.

As before, let $v \in a(U)$ be any vector not too close to the boundary of $a(U)$.

PROPOSITION 9. *Suppose there are sequences $x^{(m)}, y^{(m)} \in \mathbf{R}^n, q_m \to \infty$, such that*

$$v_m = \frac{y^{(m)} - x^{(m)}}{q_m} \to v,$$

and minimal states

$$\left(x_0^{(m)} = x^{(m)}, x_1^{(m)}, \ldots, x_{q_n-1}^{(m)}, x_{q_n}^{(m)} = y^{(m)}\right) \in \Psi_{q_n}^{x^{(m)}, y^{(m)}},$$

such that for $i = 1, \ldots, q-1$

$$\operatorname{dist}\left(x_i^{(m)}, x^{(m)} + iv^{(m)}\right) < C$$

where C is independent of m.

Then there exists a minimal orbit with rotation vector v.

Proof. Using the periodicity of the generating function we can assume that all $x_{[\frac{q_m}{2}]}^{(m)}$ lie in the unit cell. Hence, by compactness one can also assume that $x_{[\frac{q_m}{2}]}^{(m)} \to x$ and corresponding r-coordinates also converge to r. By Proposition 3 the orbit $F^k(x, r) = (x_k, r_k)$ is minimal. Fixing k, putting $i = k + \left[\frac{q_m}{2}\right]$ in (11) and letting m go to ∞ we have for sufficiently large m

$$\operatorname{dist}(x_k^{(m)}, x^{(m)} + kv) < 3C$$

and hence

$$\lim_{k \to \pm\infty} \frac{x_k - x}{k} = v.$$

$\square$

Let us discuss some of numerical results supporting the validity of the criterion of the last proposition and possible shapes of minimal orbits for irrational rotation vectors which do not produce KAM tori. The results were obtained by Mark Muldoon in his Ph.D Thesis [Mu]. Related calculations can be found in a preprint by Kook and Meiss [KM].

Muldoon studies Birkhoff minimizing periodic orbits for four-dimensional symplectic maps. The symbol $(p, q)/m$ describes such an orbit with rotation vector $\left(\frac{p}{m}, \frac{q}{m}\right)$. A typical example presented by Figure 3 displays a highly disconnected behavior for a fairly small approximation of the integrable map due to a near resonance. Nevertheless, the picture is quite far from a uniform two-dimensional Cantor set which would appear for a product of two twist maps. This is probably related to very different behavior of two positive Lyapunov exponents corresponding to the minimizing orbits. We refer to [Mu] for further discussion.

Figure 4 shows that the bounded deviation condition of Proposition 9 is highly plausible.

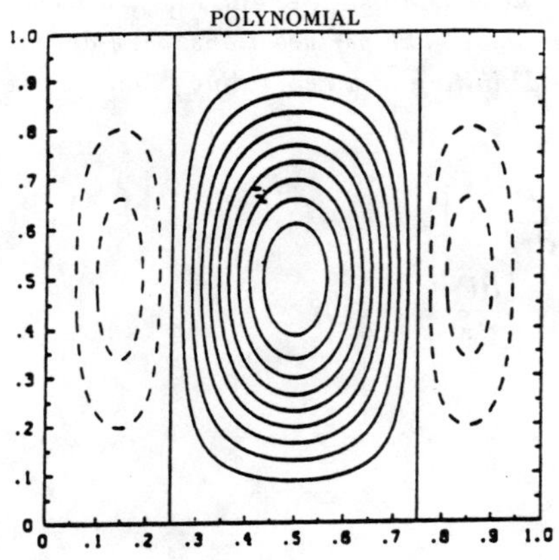

Figure 1

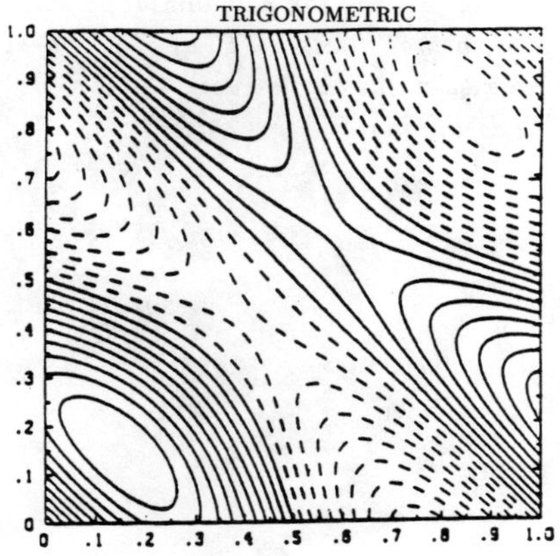

Figure 2

Pictures of the perturbations to the generating function. *We study maps generated by functions of the form*

$$H_\varepsilon(x,x') = h(x'-x) + P_\varepsilon(x,x'),$$

where

$$h(x'-x) = \frac{1}{2}\|x'-x\|^2 , \qquad P_\varepsilon(x,x') = \varepsilon P(x),$$

and

$$P(x) = \begin{cases} \textit{either} \\ P_{\text{trig}}(x) = \dfrac{1}{M_{\text{trig}}}\left\{\dfrac{1}{2}(\sin 2\pi x_0 + \sin 2\pi x_1) + \sin 2\pi(x_0+x_1)\right\}, \\ \textit{or} \\ P_{\text{poly}}(x) = \dfrac{1}{M_{\text{poly}}}\left\{\left(\left[x_0^2(1-x_0)^2(x_0-\dfrac{3}{4})\left(\dfrac{1}{4}-x_0\right)\right)\right][x_1^2(1-x_1)^2]\right\}. \end{cases}$$

The x_i, $i=0,1$ *in the formula above are real numbers, the components of the argument of the function* $P(x)$. *Call the first perturbation the trigonometric perturbation and the second the polynomial perturbation. The constants* M_{trig} *and* M_{poly} *are chosen so that* $\max_{x\in T^n} P(x) = 1$. *These are standard-like perturbations, they depend on* x, *but not on its successor,* x'. *Using the definition of a generating function one finds the map:*

$$p'(x,p) = p - \varepsilon\left[\frac{\partial P}{\partial x}(x),\right]$$
$$x'(x,p) = x - p + \varepsilon\left[\frac{\partial P}{\partial x}(x)\right].$$

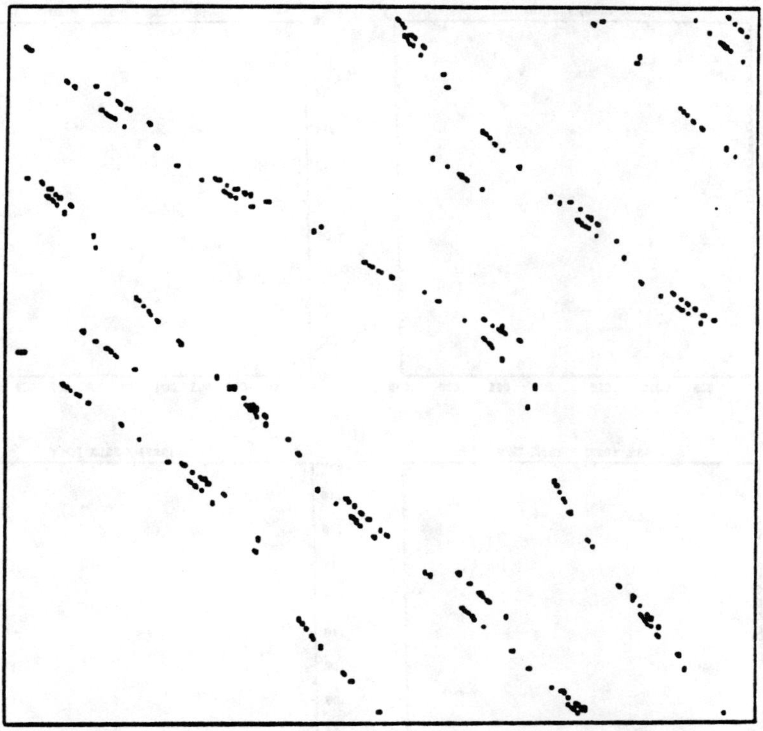

Figure 3

trigonometric perturbation
(1432, 1897)/2513

ε	0.0285
grad size	$6.259.10^{-6}$
shadow	$9.404.10^{-5}$

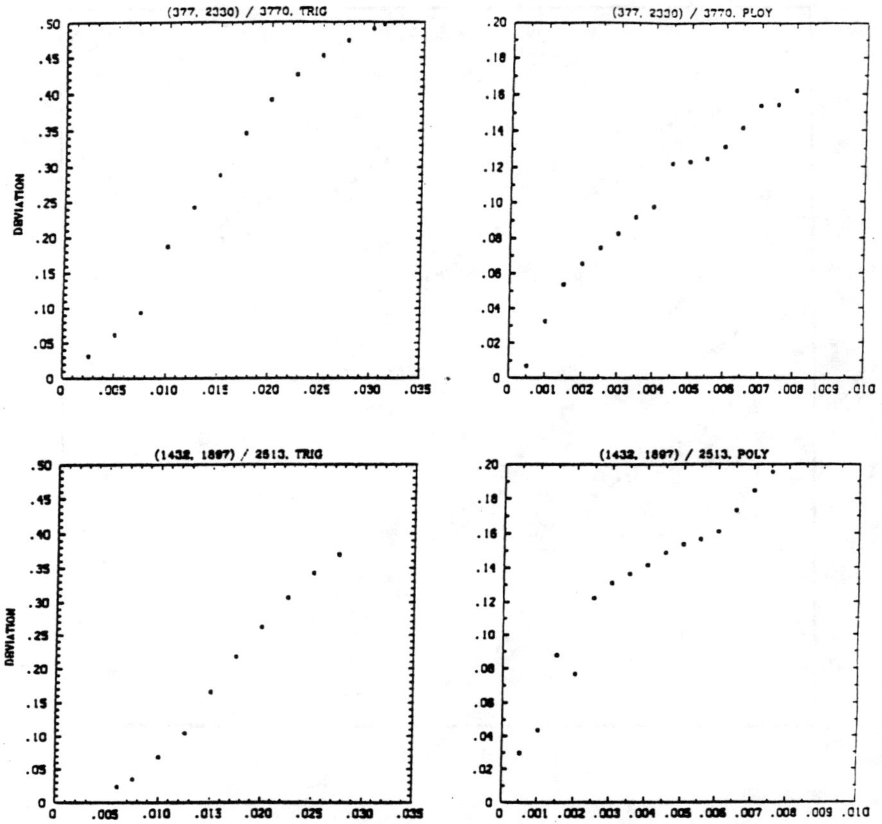

Figure 4

Deviation of points in the minimizing state from the position they would have if the state were unperturbed. The squares in the plots above represent the largest deviation found. The labels at the tops of the frames indicate the perturbation and rotation number.

REFERENCES

[A] V.I. Arnold, *Proof of A.N. Kolmogorov's theorem on the preservation of quasi-periodic motoins under small perturbations of the Hamiltonian*, Usp. Mat. Nauk SSSR 18, 5, (1963), 13–40.

[AL] S. Aubry and P.Y. Le Daeron, *The discrete Frenkel–Kontorova model and its extensions I*, Phys. D, **8** (1983), 381–422.

[B1] V. Bangert, *Mather sets for twist maps and geodesics on tori*, Dynamics Reported**1** (1987).

[B2] V. Bangert, *Minimal geodesics*, Preprint (1987).

[BK] D. Bernstein and A. Katok, *Birkhoff periodic orbits for small perturbations of completely integrable Hamiltonian systems with convex Hamiltonians*, Invent. Math. **88** (1987), 225–241.

[Bo] J.B. Bost, *Tores invariants des systèmes dynamiques hamiltonien*, séminalaire Bourbaki' Exposé **639**, Astérisque 133–134 (1986), 113–157.

[H] G.A. Hedlund, *Geodesics on two-dimensional Riemannian manifold with periodic coefficients*, Ann. of Math. **33** (1932), 719–739.

[He1] M.R. Herman, *Existence et non existence de tores invariants par des diffeomorphismes symplectiques*, Séminaire Equations aux derives partielle, Ex. XIV, 1987–88, Ecole Polytechnique.

[He2] M.R. Herman, *Inégalités a priori pour des tores langrangiens invariants par les difféomorphismes symplectiques*, Publ. Math IHES, **70** (1989), 47–101.

[K] A. Katok, *Periodic and quasi-periodic orbits for twist maps*, in Dynamical Systems and Chaos, Proceedings , Sitges 1982, Lecture Notes in Physics **179**,Springer–Verlag (1983).

[KM] H. Kook and J.D. Meiss, *Periodic orbits for reversible symplectic mappings*, Preprint (1988).

[M1] J.N. Mather, *Existence of quasi-periodic orbits for twist maps of the annulus*, Topology, 21 (1982), 457–467.

[M2] J.N. Mather, *Minimal measures*, Comm. Math. Helv. **64** (1989), 375–384.

[Mo] M. Morse, *A fundamental class of geodesics on any closed surface of genus greater than one*, Trans. Amer. Math. Soc. **26** (1924), 25–60.

[Mos1] J. Moser, *Stable and random motions in dynamical systems*, Ann. Math. Studies, **77**, (1973).

[Mos2] J. Moser, *Recent developments in the theory of Hamiltonian systems*, SIAM Review, **28** (1986), 459–485.

[Mu] M. Muldoon, *Ghosts of order in Hamiltonian systems with many degrees of freedom*, Ph.D. Thesis, Caltech (1988).